New Directions in the Philosophy of Science

Series Editor: **Steven French**, Philosophy, University of Leeds, UK

The philosophy of science is going through exciting times. New and productive relationships are being sought with the history of science. Illuminating and innovative comparisons are being developed between the philosophy of science and the philosophy of art. The role of mathematics in science is being opened up to renewed scrutiny in the light of original case studies. The philosophies of particular sciences are both drawing on, and feeding into, new work in metaphysics, and the relationships between science, metaphysics, and the philosophy of science in general are being re-examined and reconfigured.

The intention behind this new series from Palgrave Macmillan is to offer a new, dedicated publishing forum for the kind of exciting new work in the philosophy of science that embraces novel directions and fresh perspectives.

To this end, our aim is to publish books that address issues in the philosophy of science in the light of these new developments, including those that attempt to initiate a dialogue between various perspectives, offer constructive and insightful critiques, or bring new areas of science under philosophical scrutiny.

Titles include:

THE APPLICABILITY OF MATHEMATICS IN SCIENCE: INDISPENSABILITY AND ONTOLOGY
Sorin Bangu

PHILOSOPHY OF STEM CELL BIOLOGY:
Knowledge in Flesh and Blood
Melinda Bonnie Fagan

SCIENTIFIC ENQUIRY AND NATURAL KINDS:
From Planets to Mallards
P.D. Magnus

COUNTERFACTUALS AND SCIENTIFIC REALISM
Michael J. Shaffer

MODELS AS MAKE-BELIEVE:
Imagination, Fiction and Scientific Representation
Adam Toon

Forthcoming titles include:

THE PHILOSOPHY OF EPIDEMIOLOGY
Alex Broadbent

SCIENTIFIC MODELS AND REPRESENTATION
Gabriele Contessa

CAUSATION AND ITS BASIS IN FUNDAMENTAL PHYSICS
Douglas Kutach

BETWEEN SCIENCE, METAPHYSICS AND COMMON SENSE
Matteo Morganti

ARE SPECIES REAL?
Matthew Slater

THE NATURE OF CLASSIFICATION
John S. Wilkins and Malte C. Ebach

New Directions of the Philosophy of Science
Series Standing Order ISBN 978–0–230–20210–8 (hardcover)
(*outside North America only*)

You can receive future titles in this series as they are published by placing a standing order. Please contact your bookseller or, in case of difficulty, write to us at the address below with your name and address, the title of the series and the ISBN quoted above.

Customer Services Department, Macmillan Distribution Ltd, Houndmills, Basingstoke, Hampshire RG21 6XS, UK

Philosophy of Stem Cell Biology

Knowledge in Flesh and Blood

Melinda Bonnie Fagan
Rice University, USA

© Melinda Bonnie Fagan 2013

Softcover reprint of the hardcover 1st edition 2013 978-0-230-36827-9

First published 2013 by
PALGRAVE MACMILLAN

Palgrave Macmillan in the UK is an imprint of Macmillan Publishers Limited,
registered in England, company number 785998, of Houndmills, Basingstoke,
Hampshire RG21 6XS.

Palgrave Macmillan in the US is a division of St Martin's Press LLC,
175 Fifth Avenue, New York, NY 10010.

Palgrave Macmillan is the global academic imprint of the above companies
and has companies and representatives throughout the world.

Palgrave® and Macmillan® are registered trademarks in the United States,
the United Kingdom, Europe and other countries.

ISBN 978-1-349-34985-2   ISBN 978-1-137-29602-3 (eBook)
DOI 10.1057/9781137296023

This book is printed on paper suitable for recycling and made from fully
managed and sustained forest sources. Logging, pulping and manufacturing
processes are expected to conform to the environmental regulations of the
country of origin.

A catalogue record for this book is available from the British Library.

A catalog record for this book is available from the Library of Congress.

Transferred to Digital Printing in 2013

As soon as she saw this, Medea unsheathed a knife, and cut the old man's throat, and letting the old blood out, filled the dry veins with the juice. When Aeson had absorbed it, part through his mouth, and part through the wound, the white of his hair and beard quickly vanished, and a dark colour took its place. At a stroke his leanness went, and his pallor and dullness of mind. The deep hollows were filled with rounded flesh, and his limbs expanded. Aeson marvelled, recalling that this was his self of forty years ago.

– Ovid, Metamorphoses, Book 7

Biology is not destiny, but opportunity.

– Nikolas Rose, The Politics of Life Itself

Contents

List of Figures and Tables

Figures

Tables

Series Editor's Foreword

The intention behind this series is to offer a new, dedicated publishing forum for the kind of exciting new work in the philosophy of science that embraces novel directions and fresh perspectives. To this end, our aim is to publish books that address issues in the philosophy of science in the light of these new developments, including those that attempt to initiate a dialogue between various perspectives, offer constructive and insightful critiques, or bring new areas of science under philosophical scrutiny.

Melinda Fagan meets these aims with a rigorous and thought-provoking analysis of one of the most significant and challenging areas of contemporary bioscience. Stem cell research has been the focus of considerable controversy in recent years, but Fagan is primarily concerned not with the associated ethical issues, but with some of the core components of the relevant scientific practices: the status of biological entities, the role of genes in development, and, more generally, the nature and role of models, their relationship with experiment and the social epistemology of research in this area. As she says in her opening chapter, "[p]hilosophical ideas and arguments can illuminate significant features and prevailing trends in stem cell science," but "[s]tem cell research, in turn, offers valuable challenges and lessons for philosophers of science." Both philosophy and stem cell science stand to gain from bringing the two together in the fruitful and provocative way laid out in her book.

Fagan begins by setting out what she calls the "visceral phenomena of stem cell biology," presenting the basic concepts, techniques, and experiments in an accessible and engaging way. Part I of the book then analyses and clarifies the core issues using standard ideas from the philosophy of science. Part II reverses that relationship, bringing key issues in the philosophy of science to the fore and using stem cell biology to shed new light on them. The third, and final, part of the book then pulls the various strands together through a detailed consideration of the relationship between stem cell science and other fields, such as systems biology and clinical medicine.

This structure is interwoven by three crucial themes: the first is that of interaction, between cells and their environment, between cells themselves, between models of development, and between technologies and concepts from different fields. The second, related, theme is that of pluralism, in particular with regard to the models and representations deployed by scientists. Stem cell biology displays a range of abstract and concrete models that are often fluid in nature and exemplify the flux of practice. Fagan contends that a pluralistic attitude towards such representations is fundamental to understanding stem cell research. And, finally, the third theme is that of unification, offered as a counterbalance to the second, and posited as sitting at the crux of the relevant explanatory moves. In this respect context-relative results and diverse models are typically unified via collaborative interactions that illustrate how the three themes work together to characterize the interplay of unity and diversity that is a key feature of this research.

By means of this structure, and by focusing on the earlier-mentioned themes, Fagan intends for her book to itself exemplify an 'interactive epistemology' that will initiate cross-disciplinary collaboration between philosophy of science and stem cell biology. This is an ambitious and provocative aim, but one that falls precisely within the remit of the *New Directions* series. Fagan's book not only illuminates a range of core topics in the philosophy of science, and offers a skilful and hugely knowledgeable analysis of a fundamentally important field of biological research, but it also has profound implications for science policy and consideration of the social dimensions of scientific research in general. It represents an outstanding example of the 'philosophy of science in practice' that will, undoubtedly, have a major impact both within and beyond the discipline.

Steven French
Professor of Philosophy of Science
University of Leeds

Acknowledgments

This book could not have been written without the support of my colleagues and mentors, both in philosophy and biology. Particular thanks are due to Lisa Lloyd, Irv Weissman, and the Philosophy Department and Humanities Research Center at Rice – the first two for guidance at early stages of the project and the last for the resources to complete it. The ideas and arguments in this book also benefited greatly from discussions with Leo Aguila, Jordi Cat, Hasok Chang, Mike Clarke, Kim Gandy, Elihu Gerson, Richard Grandy, James Griesemer, Leonard and Leonore Herzenberg, Oleg Igoshin, Libuse Jerabek, Hanno Hock, Mi-gong Kim, Chris Kelty, Hannah Landecker, Jane Maienschein, Kirstin Matthews, Sean Morrison, Stephen Munzer, Angela Potochnik, Jutta Schickore, Paul Simmons, Jerry Spangrude, Joseph Ulatowski, Amy Wagers, Marius Wernig, and Jack Zammito. The book also benefited greatly from discussions in my Spring 2010 undergraduate course 'Perspectives on Stem Cells' (co-taught with Ann Saterbak of Bioengineering) and Fall 2010 graduate seminar in Philosophy of Science; many thanks to the participants in both. My undergraduate research assistants, Casey O'Grady, Tracey Isidro and Dandan Liu, gave invaluable help in organizing references and discussing ideas in early drafts.

Earlier versions of these arguments and ideas were presented at Indiana University (Indianapolis and Bloomington, March 2009), Rice University (November 2009), the Society for the Philosophy of Science in Practice (Minneapolis, June 2009, and Exeter, June 2011), the Center for Society and Genetics at UCLA (October 2009), the European Philosophy of Science Association (Amsterdam, October 2009), the 3rd meeting for Integrated History and Philosophy of Science (Indiana University, September 2010), the Epistemology of Modeling and Simulation Workshop (Pittsburgh, March 2011), the International Society for the History, Philosophy, and Social Studies of Biology (Salt Lake City, July 2011), the Interdisciplinarity and Systems Biology Workshop (Aarhus, August 2011), the Baker Institute for Public Policy (Houston, September 2011), and the Pacific APA (Seattle, April 2012). The last event, a focused discussion of earlier drafts of Chapters 2–4 (Part I) by biomedical experts, yielded extremely valuable feedback at a crucial time. So particular thanks is due to all participants, and Kirstin Matthews, who organized the event.

The earliest stages of research that led to this book were funded by a Doctoral Dissertation Improvement Grant from the National Science Foundation (SES-0620993) and a Dissertation Year Fellowship from the College of Arts and Sciences at Indiana University, Bloomington (2006–2007). Support for later stages was provided by a Collaborative Research Fellowship from the Humanities Research Center at Rice University (Fall 2009), three Mosle Research Fellowships (2009–2012), and a Faculty Innovation Fund grant from Rice University (2010–2012). Invaluable support in the form of scheduled leave and collegial encouragement was provided by the Rice Philosophy Department. Rice University, and its many interdisciplinary forums, offered an excellent and productive environment for this work.

Finally, many thanks to Steve French, for wonderful encouragement and support, and the editors and anonymous reviewers for Palgrave Macmillan and this series.

The cover photo of human embryonic stem cells appears with permission of Eugene Brandon, PhD and Viacyte, San Diego (March 2010).

Some of the material in Chapters 6, 8, and 9 is drawn from previously-published articles, and appears by permission of:

The MIT Press, for Melinda B. Fagan, 'Social Experiments in Stem Cell Biology', Perspectives on Science, 19:3 (Fall, 2011), 235–262 (Chapter 8)

Springer Science+Business Media, for Fagan, M.B. (2012) Waddington redux: models and explanation in stem cell and systems biology. *Biology and Philosophy* 27: 179–213 (Chapters 6 and 9).

The following Figures appear with kind permission from Elsevier Press:

Chapter 2:
Figure 1 (excerpt) of Kazutoshi Takahashi, Koji Tanabe, Mari Ohnuki, Megumi Narita, Tomoko Ichisaka, Kiichiro Tomoda, and Shinya Yamanaka, 'Induction of pluripotent stem cells from adult human fibroblasts by defined factors', *Cell* 131:5 (2007), 861–872.

Figure 4 of George Price, 'The nature of selection', *Journal of Theoretical Biology* 175:3 (1995), 389–396.

Chapter 8:
Figure 2 of Sergei Doulatov, Faiyaz Notta, Elisa Laurenti, and John Dick (2012) 'Hematopoiesis: a human perspective', *Cell Stem Cell* 10 (Feb 2012), 120–136.

Chapter 6:
Figure 1A and 1B of Qiao Zhou and Douglas Melton, 'Extreme make-over: converting one cell into another', *Cell Stem Cell* 3 (Oct 2008), 382–388.

Chapter 9:
Part of Figure 2A of Sui Huang, Ingemar Ernberg, and Stuart Kauffman, 'Cancer attractors', *Seminars in Cell & Developmental Biology* 20 (2009), 869–876.

Chapter 10:
Figure 1 of David Scadden and Alok Srivastavra, 'Advancing stem cell biology toward stem cell therapeutics', *Cell Stem Cell* 10 (Feb 2012), 149–150.

The following Figures appear with permission from John Wiley and Sons:

Chapter 4:
Figure 3A of Dov Zipori, 'The stem state: plasticity is essential, whereas self-renewal and hierarchy are optional', *Stem Cells* 23 (2005), 719–726.

Chapter 9:
Figure 2E of Sui Huang, 'Reprogramming cell fates: reconciling rarity with robustness', *BioEssays* 31 (2009), 546–560.

The following figures appear with kind permission from Springer Science+Business Media:

Chapter 6:
Figure 1 of Konrad Hochedlinger and Kathryn Plath (2009 p. 510) 'Epigenetic reprogramming and induced pluripotency', *Development* 136: 509–523, is reproduced by permission from *Development* (The Company of Biologists).

Chapter 8:
'Principles of hematopoietic stem cell biology' (2010), in M. Kondo (ed) *Hematopoietic Stem Cell Biology*, pp. 1–36, Ema, H., Kobayashi, T., and Nakauchi, H., Figure 1, New York: Humana Press.

In *Chapter 6*, Figure 2 of Shinya Yamanaka (2009) is reprinted by permission from Macmillan Publishers Ltd: *Nature*, 'Elite and stochastic models for induced pluripotent stem cell generation', *Nature* 460: 49–52.

In *Chapter 3*, part of Figure 2 of James E. Till, S. Wilson, and Ernest A. McCulloch, 'Repression of colony formation reversed by antiserum to mouse thymocytes', *Science* 169 (Sept 25, 1970), 1327–1329, is reprinted by permission from American Association for the Advancement of Science. Photo courtesy of James Till.

In *Chapter 8*, a slide illustrating a multi-laboratory study on tissue-specific stem cells is used by permission of Professors Owen Witte and Andrew Goldstein (UCLA Medical Center).

In *Chapter 10*, a poster announcing the May 2010 conference on stem cell and systems biology at UC Irvine's Center for Complex Systems is reprinted with permission of the author and Director, Arthur Lander.

Every effort has been made to trace rights holders, but if any have been inadvertently overlooked the publishers would be pleased to make the necessary arrangements at the first opportunity.

1
Visceral Phenomena

1.1 Blood and guts

This book examines stem cell biology from a philosophy of science perspective. Philosophical ideas and arguments can illuminate significant features and prevailing trends in stem cell science, examining the field in a broadly accessible way. Stem cell research, in turn, offers valuable challenges and lessons for philosophers of science. Gaining insight into this experimental, clinically-oriented field requires some departure from the traditional focus on laws and theories. Stem cell biology offers new insights on scientific models, experimental evidence, causal explanations, and social epistemology, among other topics, So both philosophy of science and stem cell biology have something to gain from their interaction. This collaborative viewpoint is defended throughout the book.

The starting point for understanding in any field is the occurrence of phenomena. Ancient astronomical theories were tasked with 'saving the phenomena' – explaining and predicting the motions of celestial bodies observable from Earth, which have furnished the subject matter of astronomy for millennia. The phenomena of stem cell biology are quite unlike the grand, remote arcs of celestial bodies. They are, instead, the stuff of *our* bodies, the mundane motions of which track our development from birth to death. This first section introduces the visceral phenomena of stem cell biology and the questions that arise from them. The next section explains how philosophy of science can help answer these questions, and introduces the basic ideas and assumptions used to do so in the chapters that follow.

Stem cell phenomena are various. Some are so familiar as to seem beneath notice: hair grows, skin is shed and replaced, cells in our bodies gradually turn over. Other species' regenerative capacities are more remarkable. Starfish and salamanders can replace severed limbs; worms and plants can re-grow an entire body from a fragment. Our own regenerative powers are demonstrated in wound-healing: bones knit, spilled blood is replaced, torn skin and muscle become smooth and whole. Most of these 'self-renewing' processes are internal and undetected. However, in the past century, innovations in 'ex vivo' cell culture have made many interior aspects of our embodied experience visible and accessible to experiment. Some of the most striking stem cell phenomena occur in transparent artificial bodies of liquid and glass. Blood, our most separable tissue, is the site of many stem cell phenomena. Blood is composed of fluid serum and cells, colloquially termed 'red' and 'white.' Red blood cells do the work of oxygen transport, while white blood cells, of which there are many types, mediate the immune response. Though these circulatory functions persist throughout life, individual blood cells do not. In humans, a single red blood cell typically circulates for a few months, then 'dies,' with its parts fragmenting to be reused or excreted. The lifespan of white blood cells is more variable, ranging from weeks to years, but the end is the same. The lifecycles of organisms and cells are not synchronized. For an organism to have a long and healthy life, new blood cells must continually develop to replace those that 'die off.'

Blood cell development is extremely sensitive to the state of the organism, including its health. Cells lost because of injury can be replaced, while immune responses defend bodily integrity by selectively amplifying cells that target particular 'invaders.' The regular movement of circulating blood in the body is interwoven with slower, but equally orchestrated, developmental movements, which maintain the balance of this tissue throughout an animal's life. The vital functions of blood require cell development and regeneration on a massive scale – about one trillion (10^{12}) new blood cells per day in adult humans. Skin and hair are similarly regulated, being tissues at the boundary of organism and environment, continually shed and replaced. The gut, another boundary area, sloughs off cells at a staggering rate: about 250 per day from each of millions of threadlike protuberances that line the small intestine. New cells rise from evocatively-termed 'crypts' in the intestinal lining, developing as they migrate into place.[1] In other organs, such as brain and ovaries, cell turnover is slow or non-existent. The challenge of stem cell biology is to explain these phenomena of cell death and renewal; specifically, how they are coordinated within and among organs and tissues to constitute a healthy, fully-developed organism.

Because all organisms begin as a single cell, the entire developmental process can be construed as a stem cell phenomenon. A fertilized egg, given appropriate nutrients and environment, undergoes repeated cell division, growth, and specialization of parts to produce a multicellular organism.[2] Stages of embryonic development in many animal species are well characterized. But their causes are not. Robustness of developmental processes across environments and species suggests a common, pre-determined 'program,' which is often attributed to genes. 'The genetic program for development' cannot be the whole story, however. All the cells of an organism's body (with minor exceptions) share the same genes. Yet, these cells exhibit different traits and perform disparate functions, both in the course of development and within a mature organism. The problem of explaining cellular diversity with invariant genes is termed the "developmental paradox" (Burian 2005). Stem cell biology approaches organismal development somewhat differently, as a cellular phenomenon. Processes by which an organism's body is constructed are conceived as cellular activities – division, differentiation, and interaction with environment – with a single point of origin, the fertilized egg. The image of development that stem cell science seeks to explain is this branching 'outward' radiation, coordinated within, and constituting, a single healthy organism.

So conceived, stem cells are the active sources of organismal development. Understanding their abilities is the key to explaining the entire process. The explanations sought are of a particular kind, giving us control over developmental processes. This brings up another basic point: stem cell biology is not 'pure science' aimed at knowledge for its own sake. The field's primary motive is to innovate new therapies for injury and disease, by harnessing cells' regenerative capacities.[3] The scope of these hoped-for therapies is very broad, precisely because the field's explanatory aims comprehend the entirety of cell development within an organism. In principle, any disorder or injury involving any cell type in the body could be ameliorated by targeted manipulation of stem cells or their products. In practice, the main clinical targets are First World diseases: cancer, diabetes, heart attack, and neurodegenerative conditions. Proponents of stem cell research anticipate "an impending revolution" both in medicine and in our ideas about cell development (Trounson 2009, xix). Perhaps the grandest and most evocative prospect is healing spinal injuries and neurodegenerative conditions, restoring the ability to walk, think, and remember.

These hopes are extravagant, easily shading into hype. What is their scientific basis? And what more do we need to know in order to realize these regenerative scenarios? The following chapters offer answers to these questions, But they do so from a philosophical perspective, rather than a scientific one. An up-to-the-minute compendium of experimental results would soon be obsolete and merely summarize what scientists themselves have said elsewhere. Instead, I focus on more enduring features of scientific method: evidence, explanation, and experiment. These are core concepts of philosophy of science.

1.2   Aims and themes

Philosophy of science originated as a method for clear communication of ideas. Stem cell biology, a relatively new field that is the focus of continual hype, misunderstanding, and argument, stands to benefit from such a treatment. Though many ethical and political aspects of stem cell research have been examined philosophically, the *science* of stem cells has not yet been articulated in an accessible and philosophically-informed way. This book aims to fill the gap. To clarify this goal, a few caveats should be noted at the outset. First, the intent is neither to reiterate scientists' own claims nor attempt to correct them. Rather, it is to provide an alternative, more general, perspective on stem cell research. This philosopher's-eye-view of the field, while neither fully comprehensive nor exhaustive, captures a number of significant features that are not so clearly resolved from the standpoint of day-to-day experience. Stem cell scientists may find this perspective useful, as an engaged, but not fully immersed, alternative to their own viewpoints. But the main goal is to make stem cell science – its methods, technology, and results – accessible to a wider audience. Philosophy of science, with its traditional emphases on methodology and evidence, augmented by new accounts of explanation, causality, scientific models, and social epistemology, has the right tools for the job.[4]

Second, this work is intended as the start of a broader discussion of philosophical issues relating to stem cell science, not the final word on the subject. What follows does not exhaust the philosophical interest of stem cell biology nor every conceivable way philosophy of science may bear on that field. Though subsequent chapters explore a number of philosophy of science topics, there is space for much future work. Of particular interest are relations between philosophy of science and ethical, political, and historical accounts of stem cell research, on which a sizable scholarly literature exists.[5] Emphasis on ethical controversies and the wider social context of stem cell science is understandable. In perhaps no other contemporary field have scientists had to be so conscious of the social context of their experimental practice, navigating a shifting landscape of political, bioethical, and financial constraints.

Yet the political, economic, and cultural significance of stem cell biology is ultimately rooted in its biomedical potential. Clarifying the experimental field that inspires the welter of expectations, concerns, hopes, and plans surrounding stem cell research will also enrich ethical, political, cultural, and historical studies. However, in what follows there is little explicit discussion of these topics. Detailed engagements with history, sociology, political philosophy, and policy-making are left for future work.

Third, the connection between stem cell biology and philosophy of science goes both ways. Stem cell biology offers valuable lessons and insights for philosophers of science. Traditionally, the central task for philosophy of science is clarification of physical theories: their structure, meaning, and connection to reality or experience. However, stem cell biology has no obvious counterpart to physical theory and bears little resemblance to 'canonical' sciences, such as Newtonian mechanics or neo-Darwinian evolutionary theory. Instead, it is driven by experiment, motivated and guided by available technology and hoped-for applications. To engage with it on its own terms presents a challenge, and a departure from traditional focus on laws and theories. Other topics in philosophy of science come to the fore, including scientific models, experimental evidence, causal explanations, and interdisciplinarity. The past decade or so has yielded new accounts of all these, which can be productively applied to stem cell biology. So rather than focusing on a single problem or debate within philosophy of science, such as realism or induction, the following chapters discuss a variety of issues, using the case of stem cell biology to critique and extend general accounts of models, experiments, and explanation. This 'topical' approach allows multiple aspects of stem cell biology to emerge more clearly in philosophical perspective.

A final caveat follows from the previous: rather than defending a single philosophical thesis, this book presents a viewpoint from which a number of interrelated philosophical claims follow. This viewpoint concerns how knowledge emerges from interaction of different elements. Though its full articulation takes place over the course of the entire book, it can be briefly sketched in terms of three themes, which recur throughout the following chapters: interaction, pluralism, and unification. I discuss each briefly, then sketch the plan of the book.

First, *interaction* is crucial, both in representations of biological development (models) and in the scientific practices that produce them (experiments). The image of development that animates stem cell biology is not that of a solitary cell that develops autonomously to reveal its inner potential. Rather, development begins with a cell interacting with its environment, whether the latter is another organism, an artificial culture, or the outside world. In any of these environments, development proceeds by cell division. Interactions among cells therefore multiply and diversify as the process moves forward, all in continual responsiveness to the environment. Our explanatory models of development should reflect this dynamic complexity. Moreover, to inform new therapies, these explanations must also indicate ways we can influence development. Successful explanations in stem cell biology not only represent biological interactions, but enable new ones: clinical interventions. Knowledge of stem cells takes the form of therapeutically-useful explanations, constrained by both these interactive aspects. Furthermore, the experiments on which such explanations are based also involve interaction – of technologies and concepts from different biomedical fields. Progress in stem cell research, as well as validation of knowledge from particular experiments, depends on collaborative interaction.

Related to the theme of interaction is *pluralism* regarding scientific representations, or models. Pluralism and emphasis on models in science go hand-in-hand. Traditionally, philosophy of science distinguished only two key domains in scientific method: theory and observation. A seminal article, *Saving the phenomena* (Bogen and Woodward 1988), proposed a third domain, associated with models. Bogen and Woodward argue that theories, exemplified by classical mechanics, explain phenomena rather than data, while data provide evidential support for phenomena. This interposes a third 'level' between theories and observational data, mediating between them. The tripartite distinction – of phenomena, explanatory theories, and observational data – allows for more nuanced accounts of scientific practice than the traditional theory: observation dichotomy. Subsequent investigations of scientific practice further support the pivotal role of models in fields ranging from physics to economics. Models in biology run the gamut from abstract to concrete. Abstract models in stem cell biology include the stem cell concept itself, as well as images of development in the form of branching hierarchies. Concrete, or material models, are striking experimental productions: "immortal" cell lines with unlimited developmental abilities (Thomson et al. 1998); embryo-like cells from "reprogrammed" adult cells (Takahashi and Yamanaka 2006); muscle, blood, and nerve tissue generated from stem cells in culture (Lanza et al. 2009, and references therein).

Experiments in stem cell biology involve the interplay of diverse abstract and concrete models. The most controversial of stem cell models – embryonic stem cells – are the material realization of a simple, idealized abstract model of organismal development. Abstractions are implicit in experimental methods used to produce concrete experimental outcomes. Concrete model organisms and model systems mingle with abstract conceptions of development and 'stemness' to yield the basic epistemic standards, goals, and organization of stem cell biology today. An approach that privileges or prioritizes one kind of model over another can yield, at best, an impoverished, and, at worst, a perniciously distorted, view of the field. Pluralism about models is fundamental for understanding, as well as practice, of stem cell research. Furthermore, new stem cell phenomena are continually being created as technologies are applied to biological materials in new ways. The field is thus open-ended and continually in flux. Experimental systems multiply and diversify, generating new phenomena. In this respect, stem cell biology resembles its subject matter: cells differentiating in local environments. Stem cell experiments, and the evidential support they provide, are relative to local contexts, which maintain their specific arrangements of technology, biomaterials, concepts, and methods. The two themes are closely related: interaction among diverse models of stem cells is basic to both scientific and philosophical understanding of the field.

The third theme, *unification*, counterbalances the diversity of models and experiments. Unification is one outcome of interaction among diverse models, and, as argued in later chapters, the crux of explanation in stem cell biology. Many aspects of stem cell science are motivated by a demand for explanatory unification of different representations. As noted earlier, cell development is conceived in terms of responsiveness to features of the environment. This context-relativity raises a number of questions. Most pointedly, can *any* cell be a stem cell, given the appropriate environment? If so, then is the idea of a 'stem *cell*' misleading? If not, then what are the limits of context-relativity in cell development? Without a general 'signature' of traits shared by stem cells across different environments, the field of stem cell biology seems fragmented, unified only by a common label for disparate objects of study. Perhaps a successor concept, such as 'stemness,' would provide a better general characterization for stem cell biology today. Parallel concerns arise for experiments and their results. Claims about stem cell capacities, insofar as they are well-supported by experimental evidence, are relative to a specific context. What connects experimental results from different contexts? On what basis can they be compared or integrated? Absent an answer, the stem cell field is just a motley collection of unlinked experiments, hardly deserving to be classified as a science.

Collaborative interactions provide a crucial part of the answer. Context-relative results from disparate sources can be integrated into a unified, comprehensive explanation of cell development by collaborative interactions across laboratories. To realize its therapeutic and explanatory goals, stem cell research must be a collaborative enterprise. It is a truism that biomedical research demands intense collaboration. The following chapters explore the implications of this idea for stem cell experiments and explanations aimed at therapeutic outcomes. Its clinical goal unifies stem cell biology providing the impetus to coordinate diverse models into unified explanations of cell development. So all three themes come together to characterize the interplay of unity and diversity in stem cell experiments, and in the science as a whole.

1.3   Plan of the book

This book both exemplifies and describes an interactive epistemology, aiming to initiate cross-disciplinary collaboration between philosophy of science and stem cell biology. The basic structure is designed to facilitate mutual engagement of the two fields. Part I focuses primarily on stem cell biology using well-established ideas from philosophy of science to clarify core issues. Part II reverses this relation, foregrounding key issues in philosophy of science and biology, and using the case of stem cell biology to engage more general arguments. Part III brings the two strands together, examining relations between stem cell biology and closely allied fields. In addition, the book imitates its subject matter in that its characterization of stem cell biology tracks a developmental arc, from simple to complex. Chapters 2–4 explicate core concepts, methods, and problems: the stem cell concept, basic design of experiments, evidential challenges, and explanatory aims. In Chapters 5–8, the picture becomes increasingly detailed and complex, with in-depth examinations of models and modeling, mechanistic explanation, experimental evidence, the role of genes in development, and collaborative organization of research. A nuanced view emerges, of stem cell research organized into diverse, yet interconnected experimental systems, with significant epistemic consequences. Chapters 9–10 extend the study still further to interfaces between stem cell biology and, respectively, systems biology and clinical medicine.

The book's overall structure thus reprises the themes that play out in its three parts. The remainder of this introduction provides a more detailed overview of chapters. Chapter 2 examines the field's most basic and central concept: the stem cell. General definitions of 'stem cell' today

converge on two developmental processes: self-renewal and differentiation. But the stem cell concept indicated by these definitions is vague
and ambiguous. Chapter 2 explicates the common core of these definitions, using two complementary strategies. The first is to construct an
abstract, inclusive, minimal model of self-renewal and differentiation.
The second is to survey concrete experimental methods used to discover
stem cells. This dual approach limns the conceptual structure common
to the various working definitions of 'stem cell,' and the shared design
of exemplary stem cell experiments. Together, the minimal stem cell
model and robust features of experiments reveal the basic organization and core concepts of the field. Chapter 3 goes more deeply into
stem cell experiments, which face two distinctive evidential challenges.
One is the 'inferential gap' between hypotheses about single cells and
experimental data about cell populations. The other is uncertainty that
follows from the fact that measurement of stem cell capacities is necessarily indirect and retrospective. Philosophical accounts of experimental
evidence help to clarify these problems and indicate how they can be
overcome. One response, proposed by a number of stem cell researchers,
is to replace the concept of 'the stem cell,' a type of cell, with 'stemness,'
a state that cells may enter or leave. Chapter 4 assesses this critical proposal, clarifying its motivation and implications. Extending arguments
by Cartwright and colleagues about theories in physics and economics,
I argue that the stemness alternative cannot replace the prevailing stem
cell concept. However, when reconciled with the latter, stemness sets
the stage for molecular explanations of stem cell capacities. This leads
into the first topic of Part II.

Chapter 5 examines mechanistic explanation – the primary mode
of explanation in stem cell biology. There is broad agreement among
philosophers of biology on the essential features of mechanisms and
the explanations that describe them. There is less consensus, however,
on what makes a mechanistic description *explanatory*. After critiquing
two prominent answers (laws and causal relations), I propose a new
account of mechanistic explanation, centered on jointly interacting
components of an overall mechanism. Chapter 6 further supports this
'joint account' by criticizing the view that genes have a privileged role
in explanations of development. Focus on stem cells, and reprogramming experiments in particular, advances this important debate. I argue
that cell reprogramming experiments do *not* support 'genetic explanatory privilege' and that models of a genetic program for gene expression
do not apply to stem cell biology. A better account of the role of genes
in development is provided by Waddington's epigenetic landscape,

updated to reflect genes' 'parity' with other components of gene expression mechanisms.

Chapter 7 turns to model organisms, the focus of most biomedical research. These concrete models are simple and tractable, shaped by epistemic and practical ends to exhibit phenomena of interest in a way accessible to observation and controlled manipulation. I argue that cultured stem cell lines have all the hallmarks of 'canonical' model organisms. This approach reveals important features of research on pluripotent stem cells, including relations among diverse model systems, the epistemic role of human embryonic stem cells, and connections between stem cells and cancer. Social dimensions of science are further explored in Chapter 8, which focuses on tissue-specific blood stem cells. A case study from the 1980s reveals social epistemic aspects of experiment, including the significance of 'experimenting communities.' More generally, the 'social experiments view' accounts for coordination of experimental results into unified explanations, with consequences for future stem cell research and policy.

Part III considers relations between stem cell biology and allied fields that are essential to its future progress, though in different ways. Chapter 9 queries the relationship between stem cell and systems biology. Systems biology is inherently interdisciplinary, aspiring to link molecular biology with mathematics, physics, engineering, and computer science. The contribution of these quantitative sciences is threefold: computer science and engineering provide instrumentation; mathematical models informed by physical chemistry yield predictions; and concepts from physics and engineering offer a starting-point for formulating 'general design principles' for living things. How, if at all, do these projects relate to stem cell biology and its experimental manipulations of concrete cells and tissues? The answer I propose builds on the joint account of mechanistic explanation (Chapter 5) and examination of Waddington's landscape (Chapter 6). The landscape model reveals, diagrammatically, the interdependence of mathematical modeling and concrete experiment in mechanistic explanations of cell development. Finally, Chapter 10 examines the relation of stem cell biology and clinical medicine. The two are inextricably linked, I argue, by the former's explicit clinical goal. Results of earlier chapters show that this therapeutic aim is constitutive for stem cell biology, shaping its normative standards for experiments, models, and explanations. This contradicts the 'value-free ideal' for science, recently criticized by proponents of 'socially relevant philosophy of science.'

A concluding section looks ahead on the road to clinical translation, noting some looming obstacles and the role philosophy can play in overcoming them. The book thus ends where it began: stem cell biology at the intersection of coordinated scientific and medical aims. But, along the way, both these goals, and the conceptual and evidential challenges that must be met to achieve them, will come more clearly into focus. Such clarification might speed our progress toward new explanations and regenerative therapies. That is the hope which animates this book.

In keeping with the themes of interaction, pluralism, and unification, each of the following chapters includes both experimental details and philosophical arguments. Readers with strong interests in one over the other may selectively skim certain sections. The main results of each chapter are summarized in a concluding section. The chapters in Part I build on one another to form a single argument and should, therefore, be read in order. Readers whose primary interest is in concepts, aims, methods, and problems of stem cell biology may want to focus on this part of the book. The chapters in Part II deal with specific topics and each stands more or less on its own. Readers with a particular interest in philosophy of biology may choose to focus on Chapters 5–8. Each chapter in Part III examines a different interdisciplinary connection, building on results from Parts I and II. So, although the final chapters can be read in isolation for their main conclusions, the full defense is cumulative.

Part I

2
Stem Cell Concepts

2.1 Introduction

2.1.1 Overview and aims

One of the first features of stem cell biology that strikes an outside observer is the sheer *variety* of stem cells: adult, embryonic, pluripotent, induced, neural, muscle, skin, blood, and so on. What 'core' stem cell concept unifies this long (and expanding) list? The straightforward answer is that a stem cell can both self-renew and give rise to other, more differentiated, cells.[6] This general definition appears in influential textbooks, journal articles, and statements by scientific organizations. For example:

(a) [A] working definition of a stem cell is a clonal, self-renewing entity that is multipotent and thus can generate several differentiated cell types.

(b) Stem cells are defined as having the capacity to both self-renew and give rise to differentiated cells.

(c) Stem cells are distinguished from other cell types by two important characteristics. First, they are unspecialized cells capable of renewing themselves through cell division, sometimes after long periods of inactivity. Second, under certain physiological or experimental conditions, they can be induced to become tissue- or organ-specific cells with special functions.

(d) Stem cell: a cell that can continuously produce unaltered daughters and also has the ability to produce daughter cells that have different, more restricted properties.[7]

These definitions converge on two cellular processes: self-renewal, production of like cells by division; and differentiation, production of specialized cells.

Yet, despite the apparent consensus, ambiguities abound. Does self-renewal involve 'continuous' production of daughter cells (d) or is it a capacity that persists through 'long periods of inactivity' (c)? Must stem cells be capable of producing multiple types of differentiated cell (a) or do cells that produce only one type of differentiated cell also qualify (b–d)? Do stem cells themselves differentiate (c) or do they produce differentiated cells through cell division (a, b, d)? On these and other points the proposed definitions do not agree. Other issues are left vague: How many cell divisions are required for cells to qualify as self-renewing – one, one hundred, or indefinitely many? Does the category 'stem cell' include only cells that *are* self-renewing, or those that *will* or *might* self-renew? And how should the category itself be understood – does it refer to a type of cell, a cell population, or a state that cells may enter? Different answers to these questions correspond to different stem cell concepts.

The aim of this chapter is to make sense of the diversity of stem cell concepts by explicating their common core. Two complementary strategies are used. The first emphasizes abstract models, the second concrete experiments. I first clarify the concepts of self-renewal and differentiation (§2.2), then combine these in a minimal, abstract model of stem cells (§2.3). Different representational assumptions and parameters, added to this minimal model, yield different stem cell concepts, which, nonetheless, share a basic structure (§2.4). The second strategy is to examine methods used to identify and characterize stem cells (§2.5). A few key protocols, spanning four decades of stem cell research, serve as exemplars, which indicate the representational assumptions and parameters prevalent in stem cell biology today. Robust features of these exemplary methods help flesh out the general stem cell concept defined by the abstract model, while their differences reveal important conceptual divisions in the field (§2.6). The result is a 'conceptual map' of stem cell biology, linking abstract definitions with experimental practices, which clarifies both its unity and diversity. I then contrast the unifying role of the abstract stem cell model with the traditional view of scientific theories (§2.7). The main conclusions of this integrative analysis, and some of their broader implications, are summarized in §2.8.

2.1.2   Stem cell basics

A brief introduction to the basic ideas, methods, and terminology of stem cell biology will set the stage. The following is a standard introductory treatment, hewing closely to those provided by scientific organizations, reviews, and prominent textbooks – also the sources of definitions such as (a)–(d) above. Readers already familiar with the basics of stem cell biology may skip ahead to §2.2, where my own analysis begins.

Stem cells are defined as unspecialized cells that can both self-renew and differentiate, that is give rise to both other stem cells and more specialized cells. They play central roles in growth, development, and maintenance of tissues and organs in multicellular organisms. Throughout an organism's life, they serve as a kind of internal repair system, replenishing the body's cells as they age and die, or are lost because of injury. Variation in the ability to regenerate, which is observed across species and developmental stages, reflects the distribution of different kinds of stem cell. Stem cells present at early stages of organismal development have the potential to produce many different kinds of specialized cell: muscle, bone, neurons, blood, etc. As development proceeds, stem cells' potential to differentiate becomes more restricted, limited to a particular organ or tissue. These stem cells are termed 'adult' or 'tissue-specific.' These distinctions are more characteristic of animals than plants; though both contain stem cells, research on the former predominates. This bias toward stem cells in animals (and, more specifically, mammals) reflects the profound influence of clinical goals on the organization and conduct of stem cell research.

All stem cells 'self-renew' by dividing. There are two modes of cell division: mitosis and meiosis. In mitosis, the genome replicates once before the cell divides. In meiosis, the genome replicates once, but two rounds of cell division follow, yielding four offspring cells with half the complement of DNA. Stem cell phenomena involve mitosis, so the term 'cell division' throughout this book refers only to that mode. When a stem cell divides, it produces two new offspring cells, each containing a copy of the parent's genome, as well as a portion of its cytoplasm, membranes, organelles, and molecular components. This process is also referred to as 'clonal proliferation.' Each offspring stem cell, like its parent, has the potential to differentiate or to remain a self-renewing stem cell. Within an organism, stem cells with a given differentiation potential maintain their numbers throughout the organism's life – a self-renewing 'stock.' Much stem cell research, however, involves placing cells in artificial culture. Artificial cell cultures consist of a transparent dish or flask of liquid media enriched with nutrients, kept at a constant (warm) temperature.[8] Cell culture reveals an important contrast between two kinds of stem cell: those that divide continuously in artificial cultures without differentiating for long time intervals and those that do not self-renew under these conditions. The former include embryonic stem cells (ESC); the latter are termed 'adult' or 'tissue-specific.' It is important to note that while ESC are derived from early-stage embryos, the term 'adult stem cell' is applied more broadly, including stem cells from fetuses and cord blood, as well as those found in mature organisms. Understanding how the balance between self-renewal and differentiation is regulated under artificial conditions and in organismal bodies is a major aim of stem cell research today.

Different types of stem cell come from different parts of an organism's body at different developmental stages. The most widely-discussed are ESC, which, as noted earlier, come from early embryos. More specifically, ESC are made from embryos at the blastocyst stage, which consist of two layers of cells: inner and outer. In normal development, the inner cell layer gives rise to embryonic tissues and, eventually, the entire organismal body. Cells from the inner layer of a blastocyst, when placed in artificial culture containing the right combination of nutrients and biochemical signals, give rise to a continuously proliferating stem cell 'line.' Finding the right culture medium for continuous self-renewal of an ESC line is a matter of trial and error. This was first accomplished for mouse cells (1981) and applied to humans nearly two decades later (1998). The human cells used in these experiments were from embryos created in vitro, as byproducts of assisted reproduction for infertile couples.

Trial-and-error is also the means of discovering culture conditions that induce stem cells to differentiate. All stem cells can differentiate to produce specialized cells with distinct structures and functions in the adult organism. But the extent of their differentiation potential varies widely. ESC have the greatest differentiation potential, being able to give rise to all types of cells in the adult organism (pluripotency). This potential is demonstrated by 'directed differentiation' in the laboratory: changing culture conditions so as to induce ESC to become neurons, muscle tissue, bone, and so on.[9] 'Differentiation in a dish' offers a window into mechanisms of development in particular tissues and organs, and also furnishes material for drug testing and models of disease. Though therapeutic hopes for these protean cells are high, clinical trials on ESC are just beginning. Successful stem cell therapies will require: (i) sufficient quantities of cells; (ii) controlled differentiation into the desired cell type; (iii) effective transplantation and engraftment into recipients, bypassing mechanisms of immune rejection; (iv) lifelong function in the host; and (v) absence of harmful side-effects, such as unregulated cell growth (tumors).

These conditions are already met by some adult stem cells, notably those that give rise to the blood and immune system. Blood-forming, or hematopoietic, stem cells (HSC) are used widely to treat leukemia and other cancers.[10] But because these adult stem cells do not self-renew continuously in culture, as ESC do, they must be found 'in situ' within organismal tissues. In most tissues and organs, stem cells are rare, comprising a small subpopulation or 'compartment.' The physiological setting of adult stem cells is referred to as their 'niche.' Creating an artificial niche in cell culture in which adult stem cells can be studied and systematically produced is an ongoing technical challenge. Within an organism, the abundance of a given type of adult stem cell at any particular time is a function of cell and tissue dynamics, the delicate interplay of cell turnover and replacement that keeps each organ functioning in conjunction with others to constitute a healthy organism. These mechanisms are still poorly understood, so identifying and using adult stem cells is important for basic, as well as clinical, research. Before the 1990s, when neural stem cells were isolated experimentally, scientists believed that no new neurons could form in the adult brain. Experimental discoveries in rats, mice, and humans, using methods pioneered in HSC, overturned that conventional wisdom.

There is long-standing controversy about the extent of differentiation potential for adult stem cells, especially those associated with blood, bone, and fat. For the most part, the 'tissue-specific' moniker seems accurate: stem cells isolated from a given organ or tissue can give rise to cells of that organ or tissue, and not others. Such cells are termed multi-, oligo-, or unipotent, depending on the complexity of the organ in question. But exceptions to tissue-specificity, especially under artificial culture conditions, are continually reported. Adult stem cells can, under at least some circumstances, be coaxed to 'transdifferentiate,' giving rise to unexpected cell types. Yet another kind of stem cell, the induced pluripotent stem cell (iPSC), is an 'engineered exception' to normal rules of development. iPSC are made by manipulating the genes and proteins of ordinary differentiated cells in culture, 'inducing' them to resemble cultured ESC. These 'reprogramming' experiments, pioneered in 2006, add a new level of complexity to our understanding of developmental pathways and cellular differentiation potential (see §2.5.1). Other important sources of stem cells include umbilical cord blood and certain types of cancer.[11] But, before going further into experimental details, I will first examine the general stem cell concept.

2.2 Consensus definition

The consensus definition of 'stem cell' involves two cellular processes: self-renewal and differentiation (see §2.1). Both need clarification.

(a) Symmetric and self-renewing

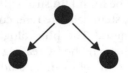

(b) Symmetric and non-self-renewing

(c) Asymmetric and self-renewing

(d) Asymmetric and non-self-renewing

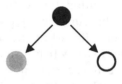

Figure 2.1　Four-way classification of cell division events, by offspring–offspring and parent–offspring comparison

2.2.1　Self-renewal

Self-renewal is a process of cell reproduction. Cells reproduce by binary division.[12] The life of a cell begins with a division event and ends with either a second division event yielding two offspring, or cell death (and no offspring). So every cell division event involves three cells: one parent and two offspring.[13] Cell division events are classified in terms of comparisons among the cells involved. Three pairwise comparisons are possible: the parent with each of its offspring and offspring with one another. By convention, classification of cell division events is based on comparison of the two offspring cells.[14] If offspring cells are the same, then the division event that produced them is termed 'symmetric' (Figure 2.1a,b); if they differ, it is 'asymmetric' (Figure 2.1c,d).

Self-renewal involves comparison of parent and offspring cells. Here, there are three possibilities: a parent cell may be the same as both, one, or none of its offspring. The first two possibilities, symmetric (Figure 2.1a) and asymmetric (Figure 2.1c), respectively, are consistent with self-renewal. The third is not, but can result from either symmetric (Figure 2.1b) or asymmetric division (Figure 2.1d). These considerations yield a four-way classification of cell division events in terms of sameness and difference within and across cell generations. But no two cells are the same in every respect. If nothing else, cells involved in a division

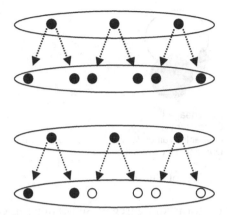

Figure 2.2 Symmetrically (top) and asymmetrically (bottom) dividing cell populations. Note that all individual cell division events shown are symmetric

event differ in position and intercellular relations. Nor are any two cells different in all respects. Cells involved in a division event share, at least, material parts and most DNA sequences. Instead, sameness and difference of cells is relative to some set of characters, such as size, shape, and concentration of a particular molecule. At a given time, each cell has some value for a given character, and it is comparisons among these values (which I shall also refer to as *traits*) that determine sameness or difference in any particular case. To summarize, cell division events can be classified by comparisons among the values of characters $C = \{x, y, z...n\}$, within and across cell generations.

This four-way classification (Figure 2.1) can also be applied to populations of dividing cells. At the population level, comparisons are of character values of parent and offspring populations (Figure 2.2). The offspring population is either homogeneous (symmetric) or heterogeneous (asymmetric) with respect to the character set of interest. In the former case, either all offspring are the same (self-renewal) or all are different from the parent population (no self-renewal). In the latter case, either some offspring are the same as the parent population (self-renewal) or none are (no self-renewal). Though parent and offspring populations are related by a set of cell division events, the population-level classification is not an aggregate of individual cell divisions so classified. Symmetric cell division events can produce any of the four population-level patterns, though asymmetric divisions can produce only asymmetrically-dividing populations. Self-renewing divisions of

Cell lineage L
Characters C = {x,y, ...n}
Character values (traits) T = {x_1..., y_1..., ... n_1...}
Duration 1≤ τ ≤∞ cell cycles

Figure 2.3   Minimal model of self-renewal

individual cells do form an aggregate self-renewing cell population, but non-self-renewing divisions of individual cells can compose a self-renewing or non-self-renewing population.[15]

Self-renewal also has a temporal aspect. For cellular processes, number of cell divisions may be more significant than calendar time, as most cell lineages in multicellular organisms die after more than 50 cell divisions. Cell cycle rate couples these two measures. Sameness across cell generations may extend from one to an indefinite number of cell cycles, or, in calendar time, from a few hours to decades. Putting these features together:

(SR1) Self-renewal occurs within cell lineage L relative to a set of characters C for duration τ, if and only if offspring cells have the same values for those characters as the parent cell(s).

Figure 2.3 depicts this general definition of self-renewal diagrammatically.

2.2.2   Differentiation

Differentiation is a core phenomenon of development: the process by which parts of a developing organism acquire diverse structural and functional traits over time. For stem cell biology, the parts of interest are cells. Every organism begins as a single cell, which, in multicellular organisms, gives rise to all the body's cells. Like self-renewal, differentiation is a comparative process involving cells of the same lineage. But, where self-renewal hinges on similarity to a parent, differentiation involves change. The simplest way to conceive of such change is in terms of a single cell with some character X (e.g. shape or size), which has value x_1 at time t_1 and x_2 at a later time t_2. But not every such change counts as differentiation. A cell that changes its character value from x_1 to x_2 thereby differentiates only if the change is in the 'direction'

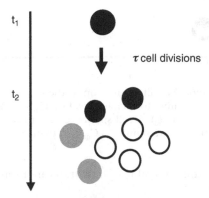

t_1

τ cell divisions

t_2

Figure 2.4 Minimal model of differentiation among contemporaneous cells of a lineage. The general process is relative to a set of characters C with alternative values (colors) and a duration (t_1–t_2), including some number $0 \leq \tau \leq \infty$ cell divisions

of more specialization and/or greater diversity. Both these comparative concepts need further discussion.

As with cell division events, we can distinguish two kinds of comparison: between contemporaneous cells in a developing lineage, and between developing and mature cells. Differentiation involves both. Cells of a developing lineage become more heterogeneous over time; they differentiate *from one another*. Such diversification can, but need not, occur via asymmetric cell division (see above). Minimally, cells in lineage L differentiate in comparison with one another over the interval t_2–t_1, relative to a set of characters C, if and only if values of C vary more at t_2 than t_1 (Figure 2.4). It is the parent–offspring lineage that undergoes differentiation in this sense, whether loosely or tightly coupled to cell division events. In general, this aspect of differentiation – diversification – is relative to a set of characters C, a set of possible alternative values for each, and a time interval t_1–t_2 including $0 \leq \tau \leq \infty$ cell divisions.

The other comparison relevant to differentiation is between cells that have completed development and cells that have not. The diverse cells composing the body of a fully-developed organism are classified according to typologies that may include hundreds of cell types. Each of the latter is defined by a cluster of character values, C_m. Differentiation in the sense of specialization involves progress toward one of these 'fates.' A cell becomes more specialized in the interval t_1–t_2 if its character values are more similar to C_m at t_2 than at t_1 (Figure 2.5). In many cases, however, there is not one cell fate to consider, but a whole array, each with a characteristic complex of traits (C_{m1}, $C_{m2}...C_{mn}$). So, in general,

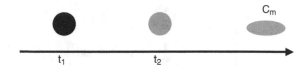

Figure 2.5   Minimal model of differentiation in the sense of cell specialization. The general process is relative to character values C_m of one or more mature cell types, and a duration (t_1–t_2). A cell specializes just in case the relevant traits (represented by color) are more similar to C_m at t_2 than at t_1

a cell specializes in the interval t_1–t_2 if its traits are more similar to some C_m at t_2 than at t_1.

The attributes of specialized mature cells are so various that it is awkward to conceive them as values of a single set of characters. For example, a neuron connects to some number of synapses and a B cell binds some antigen, but the characters of synapse number and antigen specificity are not relevant to other cell types. This idea can be expressed as an epigenetic thesis: new characters appear in the course of development, expanding the set of alternative traits and possibilities for cell variation, rather than a single character set with changing values. If this thesis is correct, then cell specialization in general is not relative to a fixed set of characters C, though any particular process will be so. A cell can become more similar to an adult cell type either by changing values of a set of characters C (x_1 to x_2), or by changing its set of characters (C to C′). In either case, the set of characters is determined primarily by the attributes of mature cells that are the end-points of the process.

Combining these two dimensions of comparison yields a general definition of cell differentiation:

(DF1) Differentiation occurs within cell lineage L during interval t_1–t_2 if and only if some cells in L change their traits such that (i) cells of L at t_2 vary more with respect to characters C than at t_1 or (ii) cells of L at t_2 have traits more similar to traits C_{m1}...C_{mk} of mature cell types {1,... k} than at t_1.

Together, the two aspects of differentiation yield a pattern that is naturally represented as a branching, tree-like structure (shown diagrammatically, for k = 2, in Figure 2.6). Though this structure corresponds to the branching diagrams representing cell division events (Figure 2.2), neither the differentiation model nor the minimal model of self-renewal represents cell division events.

To sum up, both self-renewal and differentiation involve comparison of cell traits within a lineage produced by division. As causal processes,

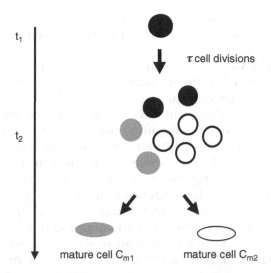

t_1

τ cell divisions

t_2

mature cell C_{m1} mature cell C_{m2}

Figure 2.6 Minimal model of cell differentiation, including both dimensions of comparison

they are opposed. Self-renewal maintains parental traits, while differentiation produces cells with diverse character values, which become progressively more like those of mature cells. The next task is to combine SR1 and DF1 in a simple, abstract description: a minimal stem cell model. Here, philosophical accounts of abstract models offer some guidance.

2.3 Models: philosophical accounts

'Model' is a notoriously ambiguous term, referring to a wide variety of representations with quite diverse forms and purposes. Philosophical accounts of scientific models and modeling can help navigate this diversity. Recent philosophical literature identifies at least three important features of abstract models in science: structure, useful representation, and mediation.[16] Model structure was first analyzed in terms of scientific theories. The influential Semantic View, pioneered by Suppes, Beth, Suppe, and Van Fraassen, defines a model as an abstract relational structure satisfying a set of postulates, theorems, or axioms used to describe a theory. A structure is formally defined as a collection of sets: of objects, relations among them, and operations.[17] But one need not identify models with mathematical or set theoretic structures to appreciate that they have a structural aspect. Construction of abstract models consists largely of specifying a set of objects and relations among them (see §2.4).

The Semantic View concerns the relation between models and theories. Subsequent philosophical studies have also examined how models relate to their representational targets – that which is modeled. The structural aspect of abstract models suggests that some form of correspondence is involved: isomorphism, partial isomorphism, or diverse relations varying case-by-case.[18] But the diverse uses of abstract models in science make it difficult to generalize about their representational relation to targets. Abstract models are used to predict empirical results, identify gaps in knowledge, demonstrate in principle possibilities, simplify complex phenomena, explicate core concepts, explain observed phenomena, and more. One general formulation that accommodates model diversity is the four-place relation: scientist S uses model X to represent target Y for purpose P.[19] The Pragmatic View associated with this schema states that the aspects of target Y that model X is to represent, as well as criteria for successful representation, are determined by model-users in light of their purposes. It follows that different models of the same target need not conflict.

A third account conceives scientific models in terms of mediation, emphasizing their 'autonomous' epistemic role in scientific practice.[20] Autonomy, or independence from theory, involves at least three distinct theses. First, models are distinct from fundamental theories (*contra* the Semantic View) and need not perfectly satisfy theoretical equations or laws. Second, theories are not the sole derivational 'drivers' of model-construction, but merely one kind of tool or resource. Third, models-as-mediators, once constructed, do not function exclusively, or even primarily, as structures for deriving predictions to test explanatory theories. Instead, they guide scientific activity in diverse ways, unexpected and unmotivated from the perspective of theory. These contributions are often grounded in models' material and phenomenal aspects. As subsequent chapters show, modeling practices in stem cell biology tend to support the Mediators View. However, philosophical discussion of models-as-mediators remains focused on the interplay of models and theories – with emphasis on *limitations* of theory. The stem cell case extends the Mediators View in a new direction (see §2.7).

These three philosophical accounts of scientific models form a sequence of progressively more inclusive approaches to science. The Semantic View concentrates on a particular kind of scientific achievement: mathematical theories. The Pragmatic View emphasizes the relevance of scientists' purposes in specifying models' representational roles. The Mediators View considers model-construction and use as part of ongoing scientific practice. This sequence also offers a useful guide for constructing a stem cell model that explicates the common core of diverse scientific definitions. I begin by specifying a structure of objects and relations. I then discuss its representational relations to targets, in light of its purpose. That purpose is to explicate the stem cell concept by exhibiting the common structure of scientists' diverse definitions of the term (§2.1). The results conform to the Mediators View, but with an experimental rather than theoretical focus.

2.4 Abstract stem cell model

2.4.1 Structure

An abstract structure is a set of objects and relations. Here, $n \geq 3$ objects are connected by two relations, termed SR and DF. In the language of directed graph theory, relations are directed edges linking nodes. So objects in the model (nodes) can be classified by the number of incoming and outgoing edges. For DF there are three cases: DF outgoing only, DF incoming only, or DF both incoming and outgoing. For SR there is only one case: SR incoming and outgoing. The model consists of the following assumptions:

(M1) Cells are bounded units that reproduce by division.
(M2) Exactly one cell has DF outgoing only.
(M3) All DF paths begin at the cell identified in (M2).
(M4) All DF paths include at least one cell with DF incoming and outgoing.
(M5) No DF edges are incoming and outgoing to the same cell.
(M6) All SR edges are incoming and outgoing to the same cell.
(M7) Each SR link has an associated duration τ.
(M8) The cell identified in (M2) has the longest τ.[21]

A consequence of these assumptions is that the model forms a hierarchy of three or more levels. Figure 2.7 shows a simple way to satisfy this structure.

This structure is highly abstract. Objects in the model (nodes) need not correspond to individual cells in biological systems, nor SF and DF to cell division events. For SR this is obvious; no biological cell divides

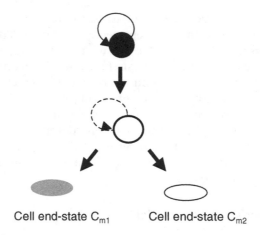

Cell end-state C_{m1} Cell end-state C_{m2}

Figure 2.7   Abstract stem cell model, combining models of self-renewal (SR) and differentiation. Longer duration of SR is depicted as a solid line, shorter duration as a broken line. Though not shown, SR is not excluded from the bottom level

to produce itself. Though objects in the model form a lineage, many characters of real cellular systems are omitted, including cell number, cell cycle number, symmetric versus asymmetric division, and parent–offspring relations. In the model, SR and DF are fully specified by (M1–M8), plus two further assumptions:

(M9)   SR produces descendants that are the same as the ancestor with respect to a set of traits C_i.

(M10)   DF produces descendants that differ from the ancestor with respect to C_i.

These last two assumptions specify the causal, reproductive aspect of SR and DF. Edges are directed from ancestors to descendants. Ancestor–descendant relations correspond to one or more parent–offspring relations. So, the model's hierarchy corresponds to biological cell lineage relationships, but at a high level of abstraction.

This abstract structure offers a minimal definition of 'stem cell.' Objects in the model are bounded units that participate in reproductive relations SR or DF as ancestors or descendants. Reproductive capacities are localized to ancestors. For a stem cell, the defining characters are just these capacities. If the characters of interest (C_i) are the capacity to participate as the ancestor in SR and the capacity to participate as the ancestor in DF, then the model's reproductive relations are reflexively specified, solely in terms of model components:

(M9′) SR produces descendants with the same capacities to engage in SR and DF as the ancestor.
(M10′) DF produces descendants with capacities to engage in SR and DF that differ from those of the ancestor.[22]

In the model's own terms, with no other traits considered, ancestors and descendants are the same if and only if they have the same capacities for producing descendants. Outgoing edges represent these capacities. Then ancestors and descendants are the same if and only if their outgoing edges link the same objects. This reproductive relation can only take the form of a loop such that a descendant occupies the same position in the hierarchy as its ancestor. In this way, self-renewal is defined by the model's structure. Differentiation is structurally defined as holding between ancestors and descendants with different capacities for producing descendants; that is, their outgoing edges differ. This relation creates the model's hierarchy: each level has a different arrangement of outgoing edges and set of descendants. Higher levels have more descendants: the object at the top level is connected to all objects in the model by outgoing edges; objects at the lowest level to none; and objects at intermediate levels to some, but not all, objects in the model. So, the model's hierarchy depicts progressively decreasing *differentiation potential*.[23]

Objects in the model can now be defined in terms of their reproductive capacities. The abstract model structurally defines a stem cell by position in a hierarchy organized by reproductive relations SR and DF. A stem cell is the unique stem of a branching structure organized by SR and DF such that each branch terminates in exactly one object. More intuitively:

- *Stem cells* are capable of long-term self-renewal and have maximum differentiation potential within the lineage.
- *Mature cells* have no differentiation potential and are capable of, at most, short-term self-renewal.
- *Progenitor cells* have less-than-maximal differentiation potential within the lineage and are capable of, at most, short-term self-renewal.

This model entails no predictions about cell phenomena, but, instead, explicates the consensus definition of 'stem cell' (§2.1). Representational assumptions are needed to link it to any biological target.

2.4.2   Representational assumptions

By making different assumptions about what its components represent, the abstract stem cell model can apply to different biological entities. Its basic structure conforms to all the working definitions (see §2.1). Comparative aspects of self-renewal and differentiation (b, c) locate stem cells in a cell lineage (d). Cells that clonally self-renew and 'can generate several differentiated cell types' fit the general definition (a), but so do 'unipotent' cells that self-renew and give rise to progenitors (c). The capacity for self-renewal may be exercised continuously (d) or intermittently (c); what matters is how long the capacity for self-renewal persists, relative to other cells in the lineage. More systematically, we can distinguish three main representational assumptions:

(R1) Objects in the model represent single cells undergoing division.
(R2) Objects in the model represent reproductively-related cell populations with statistical properties.
(R3) Objects in the model represent reproductively-related cell types.

I discuss each in turn.

A. Single cells

Perhaps the most intuitive representational assumption is that the model's objects represent single cells in biological systems and its relations, cell division events (R1). The simplest R1-model represents a single cell that divides asymmetrically to produce one offspring the same as the parent and one different, with respect to some set of traits C_i (Figure 2.1c). More complex R1-models include multiple cell division events with cell number and rate of division as variables. Any number of cell division events can be represented as a bifurcating tree diagram, with hierarchical levels corresponding to cell generations. The structure of the abstract model is preserved in any 'cell tree diagram' that satisfies two conditions: (i) ≥ 1 self-renewing division per generation and (ii) change in values of C_i. These conditions are met in many biological cases of interest. The abstract stem cell model therefore has a wide domain of application.

However, R1-models have one distinctive feature: self-renewal and differentiation are represented as realized capacities, not potentials. This induces an important structural change: self-renewal 'extends' down the hierarchy rather than being an intra-level relation; that is, in R1-models 'the stem cell' is not localized to the top of the branching hierarchy. Instead, it is a set comprised of at least one member of each cell generation: the originating parent and descendants that share the relevant character values. Oddly, then, there is no single stem cell in single-cell models of cell division. This structural peculiarity has important conceptual consequences (see below).

B. Cell populations

An alternative assumption is that objects in the abstract model represent cell populations with statistical properties derived from character values of their members (R2). In R2-models, SR and DF do not represent cell division events, but reproductive relations between cell populations (e.g. aggregates of arrows in Figure 2.2). The two relations are defined in terms of comparisons between parent and offspring populations (P and P') with respect to statistical properties; SR preserves their values, while DF does not. Individual dividing cells are represented in these models as members of P. The simplest R2-model compares cell populations across generations, omitting individual member cells and including only statistical summaries of cell properties division events in P. More complex R2-models represent cell populations and their individual members. Chapter 3 examines R2-models in more detail.

C. Cell types

A third representational assumption is that the abstract model's objects represent cell types or 'subsets' (R3).[24] Mature cells in a multicellular organism are classified into types by a cluster of character values, C_m (§2.2.2). This notion is easily extended to cells undergoing development. R3 models are thus classificatory devices, in principle applicable to any collection of developing cells. SR and DF relations in a R3 model organize cell types into a sequence of discrete stages that track the course of development, from a common 'stem' to terminally-differentiated cell types. Because biological development is a continuous process, R3 models are necessarily idealized, albeit in a way that is deeply entrenched in developmental biology.

In R3 models, SR and DF represent neither cell division events nor population-level summaries of such events, but rather reproductive relations within, and between, stages of cell development respectively. These reproductive relations remain abstract, but the typological assumption induces an asymmetry in the representational targets of SR and DF. SR represents cell reproduction within a type, which involves one or more cell division events. DF, in contrast, represents a transition between types that may or may not involve cell division. In the latter case, a cell changes its traits, thereby 'moving' to the next stage of development.

So, DF tracks cell lineage relationships, but these transitions may involve any number of cell division events, including zero.

2.4.3 Specifying parameters

The previous section shows how the abstract model's structure corresponds to a wide variety of cases, via diverse representational assumptions. But, to apply the abstract model to concrete cases, key parameters and variables must also be specified. The key parameters are precisely those features of SR1 and DF1 left out of the simpler abstract model: temporal duration and characters of interest.

A. Temporal duration

Each self-renewal relation in the abstract model has an associated duration, with stem cell self-renewal the lineage maximum. Temporal duration for cells may be measured in calendar time or number of cell divisions.[25] Calendar time, more easily measured, is the standard. In stem cell research, the duration of interest may be hours, weeks, or years. Whether a given cell counts as a stem cell or not depends, in part, on how this parameter is specified. For very short τ, there is at most one cell cycle to consider. If the cell division event is asymmetric (Figure 2.1c), the parent counts as a stem cell. In general, the shorter the duration of interest, the lower the bar to qualify as a stem cell. Most stem cell research is concerned with longer intervals, so the bar to qualify as a stem cell is higher. The duration parameter also has structural consequences: for short τ there is little variation among cells to distinguish levels of cell hierarchy. Consider a population of cells that are morphologically indistinguishable, lack specialized traits, and divide asymmetrically to give rise to offspring like themselves and more specialized offspring. Suppose that some cells in the population cycle rapidly for six months and then die, while others cycle slowly but with no limit on lifespan. For $\tau < 6$ months, both satisfy the abstract model's definition of stem cell. But, for $\tau > 6$ months, only the cells with unlimited lifespan fit the model. With longer durations, the character of self-renewal can distinguish more levels of cell hierarchy.

Relativity of stem cell definitions to duration of interest resolves a longstanding confusion about the stem–progenitor distinction. In some cases, stem cell biologists treat cells that divide rapidly and cells that can divide to produce like offspring for the length of an entire organismal lifespan as two different cell populations or types: progenitor and stem cells respectively. Others have argued that the stem–progenitor distinction is one of degree or does not exist. Experimental results are often reported in terms of 'stem/progenitor cells,' undercutting the distinction in practice. The abstract model makes sense of this equivocal situation. There is no single, absolute distinction between stem and progenitor cells. Rather, the stem cell concept is relational and relative. Whether there is a distinction between stem and progenitor cells, and, if so, where it is drawn, varies with the parameter of temporal duration.

B. Cell characters

The abstract model includes only the characters of being a bounded unit and participation in reproductive relations SR and DF. In actual cases, more characters are of interest. Specifying them yields a more realistic and detailed representation of cell lineage hierarchy. One highly variable structural feature is the number of terminating branches (n) in the hierarchy. Termini of these branches are cell fates, each distinguished by a 'signature' cluster of character values, $C_{m1}...C_{mn}$. Collectively, $C_{m1}...C_{mn}$ determine the set of characters C_i with respect to which developing cells are compared. The more terminating branches emanate from a cell, the greater its developmental potential. The maximum possible developmental potential is *totipotency*: the capacity to produce an entire organism (and, in mammals, extra-embryonic tissues) via cell division and differentiation. In animals, this capacity is limited to the fertilized egg and products of early cell divisions. In the late nineteenth and early twentieth centuries such cells were referred to as stem cells, though this terminology is not standard today.[26] Appropriately, then, the abstract model does not represent totipotency. Only 'atomized' cells of an organism are depicted, whether conceived as single cells, populations, or types. The maximum developmental potential for stem cells in the contemporary sense is *pluripotency*: ability to produce all (major) cell types of an adult organism. Somewhat more restricted stem cells are *multipotent*: able to produce some, but not all, mature cell types. Stem cells that can give rise to only a few mature cell types are *oligopotent*. The minimum differentiation potential is *unipotency*: the capacity to produce a single cell type. This qualitative classification of 'potencies' provides a framework for comparing stem cells associated with different cell traits and fates. A quantitative classification (based on n) is also possible, but this has not been put into practice.

Another variable structural feature is the number of levels separating the stem from mature cell fates. This quantity is determined by contrasts between more and less specialized cells, which define intermediate stages between the extremes of undifferentiated stem and fully-differentiated mature cells. Arrangement of branches is determined by changes in cell variation across levels. Variable characters include size, shape, internal structure, presence or absence of specific surface molecules, presence or absence of specific intracellular proteins, and presence or absence of specific behaviors (secretion, intercellular signals, movement, and so on). Specifics in a given case are determined largely by traits of terminally-differentiated cells of interest C_m. So, the number of levels tends to be roughly correlated with the number of termini (n). The greater a stem cell's developmental potential, the more elaborate the hierarchy it sits atop.

2.4.4   Application

Finally, applying the abstract model to biological cases requires a set of criteria to judge cell sameness and difference with respect to a set of characters. Here experiments and technology come to the fore. We have no access to cells except via technologies that allow us to visualize, track, or measure them. Characters attributed to cells are therefore very closely associated with methods of detection. Cells in adult organisms are distinguished by morphological, histological, and functional criteria, which figure prominently in typologies. But these are less useful for distinguishing developing cells from one another. Undifferentiated cells are often characterized *negatively*, as lacking the traits or characters of mature cells. Cell traits, fates, and technologies for distinguishing them are all closely entwined. Accordingly, specifying criteria for sameness and difference of cell character values amounts to specifying a set of methods for measuring those characters. Substantive stem cell definitions are, in this sense, relative to methods for identifying stem cells in practice. This brings us to the second strategy for explicating stem cell concepts.

2.5   Concrete methods

The history of stem cell biology is punctuated by experimental innovations.[27] A few 'exemplary methods' have been particularly influential. A survey of exemplars spanning four decades reveals several common features of stem cell experiments.

2.5.1   Induced pluripotent stem cells (2000s)

The most recent high-impact experimental innovation in stem cell biology is induction of pluripotency in fully-differentiated cells: 'direct cell reprogramming.' Lauded by *Science* magazine as 2008's *Breakthrough of the Year* and honored by the Nobel committee in 2012, the method was pioneered by Shinya Yamanaka's research team at Kyoto University (Takahashi and Yamanaka 2006). Since their 2006 paper, variations on the original protocol have yielded hundreds of new stem cell lines, publications, and drug discovery efforts. This exemplary method experimentally manipulates cells' developmental potential to produce a new kind of stem cell. Briefly: differentiated mouse or human cells are placed in artificial culture and a few (2–4) genes are added to their nuclei.[28] After several weeks, a few cells (usually ≤0.05% of the total culture) show signs of pluripotency, and are selected for further growth and proliferation. Cells that divide and self-renew under these conditions initiate a line of induced pluripotent stem cells (iPSC) which can be maintained in culture – a source of cells for other experiments.

Yamanaka and colleagues' achievement hinged on identifying iPSC as stem cells. Their original experiments had three stages: selection in original cultures, measurement of character values, and tests of differentiation potential. At each stage, experimentally-derived cells were compared with other cells. The method begins with mammalian cells in tissue culture. Initial selection was based on characters visible under the microscope: colony shape, cell size, cell shape, and nucleus/cytoplasm ratio. Differentiated skin cells formed a flat, evenly-distributed layer of fibers (Figure 2.8). But, after genes were added, a few cells showed a different morphology: round and clumped like ESC. Those of the latter type that divided to form a cell line were selected for the second stage. Characters measured were primarily molecular or biochemical (see later chapters for details).[29] Here, the important point is that the set of characters measured in the original iPSC experiments was neither exhaustive nor random. Like the morphological characters on which selection was based, the set was chosen to match character values of ESC. Cell lines with molecular characters similar to ESC went on to the third stage: tests of differentiation potential. These tests showed that some of these cell

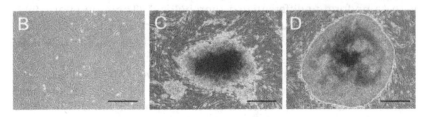

Figure 2.8 Cell cultures in iPSC experiments. Left: untreated fibroblasts (cultured skin cells). Middle: colony dissimilar to hESC. Right: hESC-like colony. Reprinted from Takahashi et al. (2007) with permission from Elsevier Press

lines are pluripotent. Pluripotency was established by placing samples of a cell line in environments that encouraged differentiation. These environments were of two types: artificial tissue culture and developing organisms. Character values of cells that appeared in these differentiation-conducive environments were then compared with those produced by ESC in the same environments. So, again, characters were chosen to match ESC experiments. But the comparisons establishing pluripotency were between *fates* of ESC and reprogrammed adult cells, with characters of normal differentiated cells providing the standard.

iPSC experiments concretely specify the key parameters of the abstract model. The duration of interest is on the order of weeks, with cells dividing continuously. The characters measured – morphological and molecular traits and fates – were chosen because of association with another stem cell type: ESC. All three representational assumptions are implicated in the iPSC method. Measured characters, such as shape and surface molecule expression, are attributes of individual cells (R1). But, manipulations are performed mainly on cell populations (R2) and determination of pluripotency hinges on similarity to more specialized cell types (R3). Overall, criteria used to identify iPSC as stem cells were based on prior experiments identifying ESC. In this sense, the iPSC method rests on an earlier exemplar.

2.5.2   Human embryonic stem cells (1990s)

Human embryonic stem cell lines (hESC) were first derived from human blastocysts by a multi-investigator group led by James Thomson of the University of Wisconsin (Thomson et al. 1998). This innovation marks the beginning of stem cell biology in its present form. The procedure begins with an early embryo in tissue culture. At this stage, embryos consist of two layers of cells. Part of the inner layer, termed the inner cell mass (ICM), is removed to a new artificial culture. In this tissue culture environment, some ICM-derived cells divide to produce colonies with "a uniform undifferentiated morphology" (Thomson et al. 1998, 1147, note 6).[30] These are selected for further culture, generating new colonies. Continuing cycles of colony formation and selection yield an 'immortal' cell line.

The experiments that identified these embryo-derived cells as stem cells had the same three stages as iPSC: selection, measurement, and test. First, cultured cells were selected for morphological similarity to cells of early human embryos *and* a previously-identified kind of stem cell, embryonal carcinoma (EC).[31] Second, molecular characters of selected cells were measured.[32] Again, the strategy was to match character values of previously-characterized stem cells, specifically those of human EC and non-human primate ESC. Some of these molecular characters (chromosome number, chromosome appearance, and telomerase activity) were also directly implicated in self-renewal. For these, character values similar to normal embryonic cells and dissimilar to cancer cells were sought. In this way, hESC were constructed to resemble normal embryonic cells. Third, differentiation potential of cells with the desired character values was tested by placing them in environments conducive to differentiation and comparing results with characters of ordinary differentiated cells. Again, environments included both artificial culture and animal bodies. But human embryos were not used in the latter. Instead, inbred mice, engineered to tolerate human transplants, provided the organismal environment.

The hESC method specifies key parameters and representational assumptions of the abstract stem cell model. The duration of interest is on the order of months or years – longer than iPSC. Otherwise, the two methods are much alike – unsurprisingly, as one is based on the other. Continuous cell division is imposed by experimental conditions. In this sense, both ESC and iPSC are constitutively self-renewing. Differentiation potential is established by comparing traits of other stem cells, cancer cells, and cells produced by normal development. As with iPSC, all three representational assumptions are implicated: characters of individual cells (R1), measurements and manipulations of cell populations (R2), and comparisons among diverse cell types (R3). However, iPSC and hESC methods differ in cell source (differentiated tissues vs early embryos) and the role of genetic manipulation (required vs optional). Their results also differ subtly – an important area of investigation for 'reprogrammers' today.

2.5.3 Blood stem cells (1970s–1980s)

Before 1998, the term 'stem cell' referred primarily to blood-making, or hematopoietic, stem cells (HSC) in mammalian bone marrow. HSC are still the focus of most stem cell research worldwide. Many other 'tissue-specific' stem cells have also been characterized: brain, gut, skin, hair, eye, testes, and muscle. But HSC have a special status. They were the first non-cancerous stem cells characterized, the first used in routine clinical practice, and they remain the best understood of all stem cells. Isolation of HSC is the exemplar for an ever-expanding catalog of ways to isolate stem cells from organs and tissues. But, unlike the previous two exemplars, HSC were not decisively established as stem cells by a single method with clear provenance. Instead, several key moments and innovations were involved, spanning several decades. The most recent such 'turning point' was in the late 1980s, involving groups from the United States and Europe.[33]

The basic method for characterizing HSC also has three stages, but departs from iPSC and ESC protocols in several important ways. One is that manipulated cells are not grown or selected in culture. HSC are already present in adult tissues; the challenge is to find them. Another difference is that developing blood cells are rather undistinguished, both morphologically and molecularly. So, *absence* of traits is an important character value. The three stages of the 1980s' method do parallel those of other exemplars, however. First, bone marrow cells were collected from an adult organism and sorted into subpopulations according to size, density, surface molecules, and cell cycle status. Second, these subpopulations were classified according to values of morphological, functional, and molecular characters, including cell size, density, and presence or absence of surface molecules. Third, differentiation potential was tested by placing samples of each subpopulation in an environment conducive to differentiation. As for other exemplary methods, these environments included both artificial cell culture and mature animals. But the latter, rather than simply undergoing development or providing a surrounding context, were also manipulated: the blood and immune system was removed by irradiation. HSC are defined as the 'subset' containing all and only the bone marrow cells that can give rise to all the major blood cell types and reconstitute the immune system. So, this exemplary method involves comparisons of organismal and cellular characters, matching results of experiments to normal blood cells and to unirradiated whole animals.

Because of the decisive role of whole animal survival, the duration of interest for HSC experiments extends at least six months in mice and, in humans, decades. Tests of differentiation potential are much like those for ESC and iPSC, with characters of interest for HSC determined by traits of blood cells of various types and stages of development. But self-renewal is established very differently being indirectly inferred at the final stage of testing rather than imposed at the outset. Another contrast is the prominence of representational assumption R3, owing to the centrality of bone marrow, subsets, or types, in the method. Yet, these types are based on cell populations extracted from bone marrow (R2), which are sorted and characterized individually (R1). So, again, all three representational assumptions are implicated. Individual cell characters, such as size, density, surface molecule expression, and cycling status, are the focus of experiments. Measurement of stem cell capacities, however, requires many cells from a given subpopulation – enough to generate visible colonies in vitro and in vivo, and to reconstitute an entire immune system for an irradiated host. The HSC method blurs the line between cell populations and cell types. The next chapter examines the implications of this feature of stem cell experiments.

2.6 Results

2.6.1 Robust framework

The three exemplary methods for identifying stem cells show several common features. All begin with cells of an organism of some species, at some stage of development. These 'source cells,' extracted from some location within the organism, give rise, under experimental conditions, to cells with capacities for self-renewal and differentiation. Each cell (or cell population or type, depending on the focus of the experiment) has a set of physical, molecular and morphological traits, which vary according to the source organism's species, and developmental stage. Differentiation potential is tested by inducing samples of a cell population to undergo development in a new context: cell culture, a mature animal, or an animal or tissue undergoing development. Results are compared with products of normal development.[34] These common features suggest a stem cell model, with three component variables: organismal origin, a 'signature' set of cell traits, and a range of cell progeny (Figure 2.9). This experimental model is grounded in concrete methods rather than abstract structures.

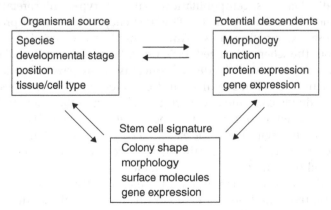

Figure 2.9 Stem cell model based on robust features of exemplary methods. Arrows denote (currently unknown) systematic relations between character values of three components

This methods-based model is, for now, merely a framework. Filling it out requires specifying relations among the values of these variables, each of which has several component characters. For the organismal source, these include species, developmental stage, and the specific site (tissue or position) from which cells are extracted. Signature character values are morphological, physical, molecular, and biochemical. Differentiated cells to which a stem cell can give rise, under defined conditions, are also distinguished by morphological and molecular characters, but with emphasis on mature cell function. Relations among values of these variables would map features of organismal source and differentiated descendants onto a 'stem cell signature,' entailing many predictions. A predictive model of this sort would describe robust relations between the values of variable characters in these three domains. We do not yet have such a model, however. 'Mapping' relations among source, signature, and progeny are largely unknown – even for the best-understood stem cells. Indeed, the 'stem cell signatures' we have are at best provisional. So the robust features of exemplary methods form a *schema*, not a predictive model. An important goal of stem cell research is to flesh out this speculative sketch.

2.6.2   Conceptual divide

Several features of the stem cell concept, as defined above, present obstacles to this goal. One is involvement of diverse representational assumptions, all compatible with the abstract stem cell model, in exemplary methods for identifying stem cells. Because these methods implicate individual cells, cell populations, and cell types, interpretation of experimental results is complex. The next chapter tackles this problem, with closer examination of experimental methods. A second obstacle to filling out the schema depicted in Figure 2.9 is a deep conceptual divide in stem cell methods. The divide concerns self-renewal. One kind of experiment, exemplified by iPSC and ESC methods, initially selects for cells that divide continuously in culture; only cells that meet this condition are tested for differentiation potential. Self-renewal is effectively imposed at the beginning of the experiment. In the other kind of experiment, exemplified by HSC methods, cells are extracted from particular tissues and their measured character values correlated to differentiation potential. Self-renewal is inferred at the end of the experiment, from long-term reconstitution of a tissue or organ in a whole animal.

This methodological contrast is significant because it goes to the core of the stem cell concept. In the first kind of experiment, self-renewal is operationally defined as continuous cell division in culture, while in the other it is the cellular process that maintains a tissue throughout an animal's lifespan. These are very different conceptions of stem cell self-renewal, which contrast both in duration and the entities taken to self-renew, i.e. their representational assumptions. In the first case, self-renewal is indefinitely prolonged and self-renewing entities are cell lines in culture.[35] Both features are consequences of cell culture methods. Cultured stem cell lines exist outside animal bodies, dividing continually until they differentiate. Self-renewal in this context is just clonal proliferation without differentiation. In the other case, self-renewal is relative to an animal's lifespan, not indefinitely open-ended. The self-renewing objects are cell populations within animal bodies, descended from injected cells that were, in turn, isolated from animal tissue. Rather than clonal proliferation, it is maintenance of an entire functional tissue that establishes self-renewal in these experiments. So, self-renewal goes hand-in-hand with differentiation, rather than being opposed to it. Though differentiation potential is tested in similar ways for both types of experiment, the two involve different *relations* of self-renewal and differentiation.

The contrast in self-renewal and its relation to differentiation is accompanied by other differences. One is extent of differentiation potential. The first kind of method is used to identify pluripotent stem cells (such as iPSC and ESC); the second to identify stem cells restricted to a particular organ or tissue (such as HSC). I will hereafter refer to these as 'pluripotency' and 'tissue-specific' methods, respectively. Pluripotent stem cells, which are created in cell culture, have a greater developmental range than stem cells that give rise to particular tissues and organs.[36] Contrast in differentiation potential, however, does not mark a fundamental conceptual division in stem cell biology. Stem cells can be ordered in a single hierarchical framework, with higher levels including lower ones: pluripotent, multipotent, and unipotent. It is the way differentiation potential is related to self-renewal that 'bifurcates' stem cell biology into two distinct branches of research. Tissue-specific methods assess self-renewal and differentiation together. Pluripotent methods assess differentiation potential of cell lines selected for self-renewal.

This bifurcation is reflected in the politically-charged division of stem cell biology into two branches: 'adult' and 'embryonic.' These widely-used labels coincide roughly with the two classes of methods distinguished above, but emphasize the stage of the source organism. Adult stem cell research is focused on stem cells present in adult tissues; ESC research is focused on cells derived from early embryos. iPSC fall into neither category, though their similarities with ESC affiliate them with the embryonic branch. ESC research aims to identify and characterize pluripotent stem cells, as well as the mechanisms by which they give rise to different cell types. Animal bodies come into this research only at the final stage, testing differentiation. In contrast, adult stem cell research aims to identify and characterize tissue-specific stem cells in blood, brain, heart, liver, skin, hair, gut, kidney, and so on. This approach presupposes a developed organism, with differentiated organs and tissues. Adult stem cells are thought to be normally tissue-specific and, therefore, multi-, oligo-, or unipotent, depending on the complexity of the tissue in question. HSC, for example, are multipotent: they can divide to produce all types of blood cells (including cells of the immune system). Muscle stem cells are oligopotent, as there are only a few types of muscle cells. Whether stem cells isolated from adult tissue have greater differentiation potential than their location suggests has long been debated.[37] But the adult/embryonic distinction, as such, is not fundamental for stem cell biology. The age of the source organism is just one variable among many in stem cell experiments. Contrasting concepts of self-renewal in relation to differentiation are more significant.

2.7   Models and theories

2.7.1   Unity and diversity

The two approaches of this chapter can now be brought together to explicate the unity and diversity of the stem cell concept. Its unity is demonstrated by the abstract stem cell model. This model defines a stem cell by position in a cell hierarchy organized by reproductive relations (§2.4.1). Many diverse cases can be fit to this simple structure. 'Fit' is accomplished by making representational assumptions about its objects and relations, and specifying key parameters and variables. Different assumptions and parameters yield different substantive stem cell concepts. So what counts as a stem cell depends on how the basic stem cell concept is fleshed out in experimental practice. This result accounts for the unity and diversity of the stem cell concept today.

Exemplary methods for identifying stem cells, though diverse, exhibit a few robust features. A general schema for systematically relating traits of organismal context, stem cells, and differentiated progeny identifies a shared goal for stem cell experiments (Figure 2.9). But, as we do not yet know how to fill in this schema, this 'model in the methods' is only an outline, rather than a unified, biologically substantive stem cell concept. For now, substantive stem cell concepts are diverse, relative to multiple parameters: experimental methods, organismal sources, and traits and durations of interest. Unless these parameters are specified, the term 'stem cell' is deeply ambiguous. Moreover, exemplary stem cell methods involve multiple representational assumptions, implicating single cells, cell populations, and cell types. Interpretation of experimental results in terms of biological entities needs further investigation, which is undertaken in Chapter 3.

2.7.2 Price's equation[38]

Diversity of models is often contrasted with unity of theories. Yet, unifying theories (and their compatriots, general laws) have been conspicuously absent from the discussion so far. The abstract stem cell model, though it plays a unificatory role, is not a theory in the traditional sense. The difference is illustrated by an example from another area of biology: George Price's 'general model of selection' (1995). Price's model is general in the same sense as the abstract stem cell model and has a similar purpose. It represents a causal process to which all cases of selection, biological and otherwise, correspond (Figure 2.10). The process involves two populations (P and P′) subdivided into one-to-one corresponding 'packages' (p_i and $p_i′$). Each package has some quantity (w_i, $w_i′$) of something with a specific character value (x_i, $x_i′$). For example, each beaker (p_i) in a set (P) contains a volume (w_i) of fluid at concentration (x_i). P′ is produced by pouring some volume of fluid from each beaker in P into a second beaker – a causal 'derivation' satisfying the requirement of one-to-one correspondence. In this model, selection on x in P is defined as the process of producing P′ from P such that 'offspring' amounts $w_i′$ are systematically related to 'parent' character values x_i.

Price's model defines a concept to which different kinds of selection (natural, sexual, artificial, intentional, multilevel, and so on) correspond as special cases, unified under a single general description.[39] This description is simple, non-mathematical, and represented as an abstract structure of objects and relations. Similarly, the abstract stem cell model defines a concept to which all kinds of stem cell correspond as special cases. It, too, is a simple, non-mathematical structure, including all and only the features of self-renewal, differentiation, and their relation needed to unify disparate cases.

Price's general model of selection is closely related to his eponymous equation:

$$\Delta X\bar{w} = \text{Cov}(w_i, x_i) + \text{Exp}(w_i \Delta x_i)$$

This equation states that the average character value of a trait before and after evolutionary change (weighted by average fitness) is equal to the

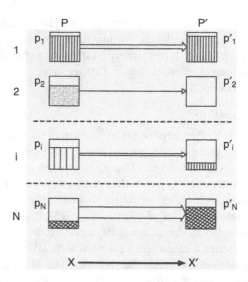

Figure 2.10   Price's general selection model. Each package has some value x_i or x_i' for character x, represented by intensity of shading. That which bears character x is present in some quantity (w_i or w_i'), represented by the area shaded. The thickness of arrows connecting the two populations represents the fractional amount selected (w_i'/w_i). Line 1 shows corresponding packages with the same amounts and numerical values. Line 2 shows corresponding packages where p_i' is empty; $w_i' = 0$. Line i shows property change: $x_i \neq x_i'$, where the amounts w_i and w_i' also differ ($w_i > w_i'$). Line N shows a case in which properties in corresponding packages (N, N') are the same, but amounts differ in the other direction ($w_i < w_i'$). The relation $X \rightarrow X'$ at the bottom represents the change in some population property X related to property x of individual members (e.g. the mean). Reprinted from Price (1995) with permission from Elsevier Press

sum of a covariance and an expectation or average of fitness times the difference in character value between parent and offspring.[40] The Price equation, arguably "the closest anyone has come to a general, abstract theory of selection" (Frank 1995, 380), exhibits features traditionally associated with scientific theories. It is a mathematical tautology that exactly partitions statistical components of evolutionary change, whose terms can be further specified to cover many diverse cases, corresponding to necessary relations among measurable quantities. In contrast, Price's general *model* is a schematic diagram representing a particular type of causal process. Rather than a necessary formal relation, it is a proposed definition, grounded not in mathematical rules but by its ability to unify disparate cases and thereby clarify a common underlying concept.

Though Price's contributions to evolutionary biology do not, of course, exhaust the meanings of 'model' and 'theory,' they do clearly illustrate the contrast between *one* sort of model and *one* sort of theory, both of which have important roles in science. For Price, the theory guides construction of his model. Properties common to all selection processes do not simply emerge from careful consideration of a simple case (the 'beakers' example). Instead, Price appeals to the terms of his eponymous equation: correlation (covariance) of the value of a trait and fitness among members of a population.[41] His equation identifies as crucial the relation between concentration and the amount of solution transferred. The properties included in the general model are exactly those used to measure selection, as defined in the Price equation. In this sense, the two define the same concept.

2.7.3 Models without theories

In stem cell biology today, there is no counterpart to the Price equation.[42] In this sense, the field lacks theories. But the abstract stem cell model does play a background unifying role, which this chapter aims to make explicit. Instead of formal theory, this model is constructed in accordance with minimal biological assumptions and connected to biological targets via concrete experimental methods. To say that stem cell biology lacks theories, in the Price equation sense, is not to say that the field lacks conceptual content, testable hypotheses, or abstract models. All three are pervasive. My claim is about theories of the sort exemplified by the Price equation: abstract formalisms that hold generally, necessarily, or universally. Theories in this sense do not play a significant role in stem cell biology today. Instead, the field is epistemically organized around exemplary methods and model systems. This situation could, of course, change – sciences are not immutable. But, for now, experiment rather than theory prevails.

2.8 Conclusions

The main results of this chapter are as follows. The abstract stem cell model, though not a theory in the traditional sense, is a minimal unifying framework for stem cell biology. In this model, a stem cell is defined as the unique stem of a lineage consisting of one or more branches, each with a distinct terminus. Termini represent a stem cell's differentiation potential. Within a cell lineage hierarchy, a stem cell has maximal self-renewal and differentiation potential. Stem cells so defined are relative to cell lineage, a set of characters, and a temporal duration of interest.

Table 2.1   Main foci of stem cell research today, classified in terms of experimental schema. Within each type, different time intervals, characters, and methods for determining sameness and difference may be used

Stem cell	Source	Signature	Descendants
hESC	Human blastocyst inner cell mass embryonic	Unknown	All three germ layers in vitro (pluripotent)
mESC	Mouse blastocyst inner cell mass embryonic	Unknown	All three germ layers in vitro, germ cells, whole animal (pluripotent)
hHSC	Human juvenile or adult bone marrow, umbilical cord blood, or peripheral blood	Various (CD34+)	Blood, immune system, others?
mHSC	Mouse juvenile or adult bone marrow, peripheral blood, spleen	Various (Lin-Sca-1+)	Blood, immune system
NSC	Mouse, rat, human adult or embryonic brain, nerves	Various	Neurons (in vitro, in vivo)
h-iPSC	Human embryonic or adult various	Unknown	All three germ layers in vitro (pluripotent)
m-iPSC	Mouse embryonic or adult various	Unknown	All three germ layers in vitro, germ cells, whole animal (pluripotent)
epiSC	Mouse blastocyst (late) inner cell mass embryonic	Unknown (≈hESC?)	All three germ layers in vitro
GSC	Human later embryo genital ridge	Unknown	All three germ layers in vitro

So the basic stem cell concept – the shared core of diverse working definitions such as (a–d) – is relational and relative. The abstract model can be elaborated in diverse ways, using different representational assumptions and parameters. Experimental methods for identifying stem cells specify parameters and representational assumptions that the minimal model leaves undetermined. In this way, the abstract model is complemented by concrete experimental methods. Another way to state this idea is that experimental methods implicitly model stem cell concepts.

Exemplary methods for identifying stem cells share a robust pattern: remove cells from an organismal source, place them in a context in which their traits can be measured, then move cells to another environmental context to measure stem cell capacities. These methods involve at least three different representational assumptions, concerning, respectively, cell types, single cells, and cell populations. Different kinds of stem cell are distinguished by organismal source, a set of 'signature' character values, duration of self-renewal and scope of differentiation potential. Exemplary methods thus imply a robust, substantive stem cell framework (Figure 2.9). Results of a stem cell experiment map correlations of character values between an organismal source, extracted cells, and differentiated cells to which the former give rise, under controlled conditions. Generalizations in stem cell biology could take the form of robust 'mapping relations' between these three sets of variables. Currently, however, this experimental schema is only partially filled out. Table 2.1 uses this schema to summarize the main varieties of stem cell that are currently being studied.

These results have several broader implications. For one, stem cell biology does not need theories to be a successful and important science. Absence of formal theories does not make a field something less than a science, but an experimental science. Stem cell biology is not a 'protoscience' awaiting a theory, at which point philosophers might deign to examine it in the traditional framework. Rather, it is a science without theories, organized instead around experiments and models. Within this organization, there is a deep conceptual divide. Stem cell experiments fall into two main groups, according to the way self-renewal is conceived. In one kind of experiment, self-renewal is operationalized as continuous cell division in culture, while, for the other, it is a cellular process that maintains a tissue throughout an animal's lifespan. The former involve pluripotent, the latter tissue-specific, stem cells. This distinction maps roughly onto the more familiar embryonic-adult distinction. But the age of source organism does not correspond to a deep conceptual divide. Emphasis on the embryonic-adult distinction as such is, therefore, something of a red herring. This suggests that the distinction serves political rather than scientific purposes. Moreover, as later chapters argue, the adult/embryonic distinction works against progress in stem cell research. The pluripotent/tissue-specific distinction is preferable.

3
Don't Know What You've Got 'Til It's Gone: Evidence in Stem Cell Experiments

3.1 Structure of experiments

As Chapter 2 showed, exemplary methods for identifying stem cells share a basic structure of three stages (Figure 3.1). The starting point is a multicellular organism – the source of cells. From this source, cells are extracted and the values of some of their characters measured. These cells (or a sample thereof) are then manipulated so as to realize capacities for self-renewal and differentiation. Each experiment involves two manipulations. In the first, cells are removed from their original organismal context and placed in a new environment in which their traits can be measured. In the second, measured cells are transferred to yet a third environmental context, which allows stem cell capacities to be realized. Finally, the amount of self-renewal and differentiation is measured. Stem cell experiments[43] thus consist of two manipulations, each followed by measurements. The objects manipulated are cells from some organismal source.

This basic method identifies stem cells by three sets of characters: of organismal source, of extracted cells, and of progeny cells (see §2.6). The characters included in the first and third sets are standardized and robust across a wide range of experiments. For organismal source, these characters are species, developmental stage, and tissue or position within the organism.[44] Character values for the organismal source are determined by choice of materials for an experiment: mouse or human; embryonic or adult; blood, muscle, or a region of the early embryo. Character values in the other two sets are measured during an experiment and comprise its results. For progeny cells, these characters include attributes of mature cell types: morphology, expression of specific genes and proteins, and function within an organism. Exactly which characters comprise the set

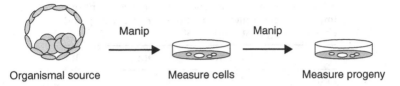

Organismal source   Measure cells   Measure progeny

Figure 3.1   Basic design of experiments aimed at isolating and characterizing stem cells

depends on the type of differentiated cells expected. For blood cells, the relevant characters are associated with immune function; for neurons, electrochemical function; for germ cells, morphological and genetic traits of gametes. Though the set of characters varies across experiments, for any particular experiment the characters of interest are established in advance: part of the standard set of morphological, biochemical, and functional traits used to classify cells in multicellular organisms.

In contrast, there are no such pre-established criteria for inclusion in the set of characters of extracted cells, i.e. presumptive stem cells. These characters vary widely across experiments, shifting rapidly in response to technical innovations and new results within the field. Yet, measurement of their values is the linchpin of stem cell experiments. Experiments aimed at isolating and characterizing stem cells unequivocally succeed just in case they reveal the 'signature' traits of stem cells from a given source. If this is impossible, then the central research program of stem cell biology is off-track and requires redirection. There are at least two ways the prevailing research program might be doomed to failure. The first is an insurmountable evidential gap between data produced by stem cell experiments and hypotheses about signature traits, such that the former cannot support the latter. The second is an insurmountable evidential gap between data about progeny cells and hypotheses about stem cell capacities, such that the former underdetermine the latter. The structure of stem cell experiments gives reason for concern about each possibility. To see this, a closer look at experimental procedures is required.

3.2   Exemplars revisited

Chapter 2 showed that stem cell experiments involve multiple representational assumptions, implicating single cells (R1), cell populations (R2), and cell types (R3). Here, I show how these different assumptions figure in the structure of two exemplary methods: one from each of the two

Table 3.1 Summary of representational assumptions implicit in experimental methods for identifying and characterizing hESC and mHSC

	R1	R2	R3
hESC			
Manip 1	X	√	X
Meas 1	√	√	X
Manip 2	X	√	X
Meas 2	√	√	√
mHSC			
Manip 1	X	√	X
Meas 1	√	√	√
Manip 2	X	√	X
Meas 2	√	√	√

main branches of stem cell research.[45] Table 3.1 breaks down the role of each representational assumption by experimental stage.

3.2.1   Human embryonic stem cells

For human embryonic stem cells (hESC), the first manipulation is to extract the inner layer from an early human embryo and place it in tissue culture. If the culture is successful, some of the extracted cells divide to produce 'outgrowth', which is collected, mechanically dissociated into 'clumps' of 50–100 cells, and moved to a fresh culture. Clumps that contain rapidly-dividing cells expand to form colonies. Continuation of this cycle of selection and growth constitutes a human embryonic cell line. The objects manipulated at this stage are cell populations (R2) with a common origin: an embryo or cell colony. Once the cell line is established, values are measured for chromosome number and appearance, expression of specific molecules on the cell surface, activity of certain intracellular proteins, expression of certain genes, and, more recently, global gene expression and epigenetic modifications. All these characters are attributes of a single cell (R1; Figure 3.2). But measuring them requires taking samples of a cell line and subjecting these samples to various treatments.[46] These procedures are performed on cell populations (R2) – aliquots of a growing cell line – rather than single cells.

In the second manipulation, samples of a cell line are transferred to an environment that encourages differentiation: cell culture or an animal

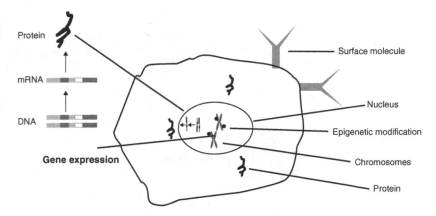

Figure 3.2 Cellular location of characters measured in stem cell experiments

body. By varying cell culture conditions, differentiation can be biased toward a particular cell type, such as neurons, cardiac muscle, or blood (Figure 3.3a). Samples of the same cell line placed in a range of different culture environments collectively reveal that cell line's developmental potential. Alternatively, cells in artificial culture can be induced to differentiate by changing their spatial arrangement. When allowed to overgrow one another, cultured mammalian embryonic cells form 'embryoid bodies': loosely-organized spheres consisting of different cell types (Figure 3.3b). Finally, if injected into immunodeficient mice, samples of a human embryonic line produce tumors that contain many human cell types ('teratomas'; Figure 3.3c).[47] Again, the objects of experimental manipulation in each case are cell populations – clumps of cells are injected into animals or transferred between culture conditions.

The second measurements are of differentiated cultured cells, embryoid bodies, and teratomas, and, as in the first set, yield values of single cell characters: cell morphology, surface molecules, specific molecules involved in gene expression, and physiological function (R1). Tests for these characters, as in the previous step, require cell populations (R2). Moreover, the significance of these measurements is to assign progeny cells to diverse cell types, representing all three germ layers and/or a wide range of tissues (R3). Differentiation potential is measured by determining the range of different cell types produced by transferred cells under experimental conditions. Self-renewal, as discussed in Chapter 2, is imposed at the outset of the experiment.

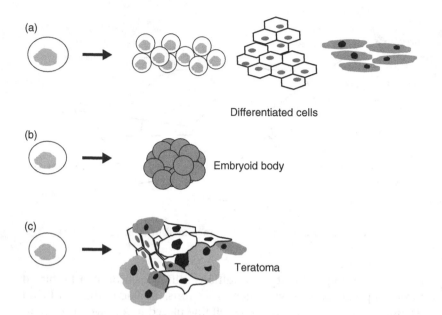

Differentiated cells

Embryoid body

Teratoma

Figure 3.3   Methods of measuring hESC differentiation potential. (a) Directed *in vitro* differentiation to multiple cell types. (b) Embryoid bodies in cell culture. (c) Teratomas in immunodeficient mice

3.2.2   Mouse blood stem cells

The first manipulation in mouse blood stem cell (mHSC) experiments is to remove cells from adult mouse bone marrow, which contains many cell types in close contact with one another.[48] Cells are collected by flushing the interior of the femur with saline, yielding a mixed cell suspension. The suspended cells are then measured and sorted at the single-cell level by a crucial piece of technology: fluorescence-activated cell sorting (FACS). FACS separates a population of cells into discrete 'compartments' or 'subsets' on the basis of quantitative traits: size, density, and number of cell surface molecules. Fluorescently-labeled cells pass one at a time through an electrostatic field which differentially deflects them based on light scatter (indicating cell size and density) and degree of fluorescence (indicating specific surface molecules). This procedure involves all three representational assumptions: single cells are measured (R1) and subdivided into populations for further use (R2) using combinations of character values that define cell types or 'compartments' (R3).

In the second manipulation, samples of each sorted compartment are transferred to an environment that encourages differentiation (R2).

As with hESC, cell populations are transferred to both artificial culture and whole animals – environments engineered to encourage differentiation toward particular cell types. For mHSCs, these are blood cell types, of which there are many. The second measurements are of characters indicating cell differentiation and self-renewal. Tissue culture measurements of HSC differentiation are very like those for ESC (R1, R2, R3), though geared to a narrower range of cell types. Whole-animal tests, however, are quite different. Sorted bone marrow cell populations are transferred to mice (of the same strain) whose blood and immune cells have been removed by radiation. Doses that do not obviously damage other tissues cause death by infection or hemorrhage within 14 days. Selective removal of these cells creates a physiological 'space' for transferred cells to fill and a clear criterion of stem cell function: irradiated mice with bone marrow cell transplants either survive or not. Long-term survival indicates both differentiation and self-renewal capacities of injected cells.

3.3 Cells and populations

A natural interpretation of experiments with the above design is that they test hypotheses about a distinctive theoretical entity: 'the stem cell.' The characters measured are attributes of single cells. Each cell has a shape, size, and nucleus-to-cytoplasm ratio. The morphological traits used to classify cell types – whether mature or undergoing development – are features of individual cells. Cell surface molecules are, evidently, expressed on a cell's outer membrane; gene and protein expression result from intracellular processes such as transcription, translation, multiple levels of RNA processing, protein packaging, and transport. 'Global' gene expression provides a 'snapshot' of the genes actively being transcribed in a cell's nucleus. Epigenetic modifications are chemical changes to proteins associated with chromosomes or to nuclear DNA itself. All these molecular entities are localized to individual cells (Figure 3.2). Moreover, stem cell scientists are clearly interested in identifying the characteristics, or 'signature,' of individual stem cells. In the early twenty-first century, laboratories at Harvard and Princeton pursued "the possibility of defining a stem cell by its constellation of active genes" (Fortunel et al. 2003, 393b) – with controversial results (see below). Ongoing research programs seek determinants of 'embryonic stem cell identity' in molecules localized to the nucleus of individual cells. These efforts are examined in later chapters; for now, the key point is that hypotheses about single stem cells are of interest to researchers.

However, with the exception of FACS, none of the methods discussed above measure single cells. Rather, what is measured, and manipulated, are 'clumps,' or cell populations. Tellingly, "none of the [original] ES cell lines was derived by clonal expansion of a single cell" (Thomson et al. 1998, 1146). Radiation rescue requires about 10^6–10^7 normal bone marrow cells and least 2×10^2 cells of more 'enriched' mHSC subpopulations.[49] Data about molecular traits requires detectable amounts of specific molecules, obtained by pooling many cells (typically $\geq 10^3$) together then breaking them apart to isolate the molecular components of interest. Microscopic observations of cell shape and structure, though made on individual cells, are aggregated to yield a population-level result, such as 'strongly positive' (>70% cells fluoresce) or negative (<10% cells fluoresce).[50] Though FACS does measure character values for single cells, stem cell experiments do not work by this technology alone (see Chapter 8). So there is an evidential gap between data and single-cell hypotheses in the exemplary methods described above. This discontinuity, between measurement and manipulation of cell populations, and hypotheses about traits of individual cells, is a general feature of stem cell research. This raises an important evidential question: Can data obtained from cell populations support hypotheses about single cells?

3.3.1   General form of the problem

This question can be precisely addressed in a general framework. In philosophy of science, evidence and hypothesis are conceived as sentences with a certain probability of being true; e and H respectively. Alternative hypotheses are distinguished by subscripts: H_1, H_2, etc. The hypotheses of concern here attribute character values to single cells. Let g_1 be a value of character G, P a cell population identified and characterized by a stem cell experiment, and S a member of P.[51] Two alternative hypotheses are:

(H_s) cell S has g_1.
(H_p) cells in population P have g_1.

Note that the two are alternatives in the sense of being different hypotheses, but are not mutually exclusive. Let e_g be data from the experiment – the results of measuring character G. Though the experiment involves two sets of measurements, the issue is most significant for the first, which aims to measure a stem cell signature. Successful experiments provide strong evidence for H_p.[52] The question is whether they can also do so for H_s.

This situation can be represented as a schematic inductive argument for H_s:

P1 Stem cell experiment M isolates cell population P from organismal source OS.

P2 M reliably measures character G in P, yielding data e_g.

P3 e_g strongly supports H_p (Some cells in P have g_1).

C H_s is true (cell S has g_1).

The question is whether premises P1–P3 provide adequate (or any) support for C and, if not, what further assumptions are needed. Philosophical accounts of evidence–hypothesis relations provide guidance here.

3.3.2 Philosophical accounts

Three prominent accounts of evidence and hypothesis deliver similar verdicts on this case: Bayesianism, likelihoodism, and error statistics. Though not exhaustive, these three represent the main philosophical theories of evidence relevant for this example. Bayesianism defines evidential support (confirmation) in terms of probability theory, such that evidence e confirms hypothesis H if and only if $Pr(H|e) > Pr(H)$. Evidential support is the difference that evidence makes to the probability that the hypothesis is true, allowing precise comparison of the degree of support for alternative hypotheses by a given body of evidence.[53] Bayes' theorem allows one to calculate $Pr(H|e)$ – the probability of H conditional on (or given) e – in terms of the total probability of the evidence $Pr(e)$, prior probability of the hypothesis $Pr(H)$, and likelihood of the hypothesis $Pr(e|H)$. Though exact prior probability values cannot be assigned to H_s and H_p, necessarily $Pr(H_p) \geq Pr(H_s)$, as a hypothesis about a particular member of a population is logically stronger than a hypothesis about some members of that population. Relative to the cell lineage of interest it is reasonable to assume that $Pr(H_s)$ and $Pr(H_p)$ are low. If most cells have g_1, this character value would not be a plausible stem cell signature. By assumption, e_g strongly supports H_p, so $Pr(e_g|H_p)$ is very high, and $Pr(e_g|\neg H_p)$ very low.[54]

The situation for H_s is less clear, precisely because stem cell experiments do not measure characters of individual cells. For molecular traits, in particular, what is measured is not a population statistic (such as a mean character value for G) but an aggregate result obtained by pooling many individual cells, often by destroying their boundaries and organization. Without information about the distribution of values of

G in P, there are no empirical grounds for estimating $Pr(e_g|H_s)$. If $Pr(e_g|H_p) \approx Pr(e_g|H_s) \gg Pr(e_g|\neg H_s)$, then the degree of confirmation for H_s and H_p is similar. Though Bayesianism allows for this possibility, there is no obvious justification for this assumption. One might also suppose $Pr(e_g|H_s) \approx Pr(e_g|\neg H_s) \approx Pr(e_g)$, from which it follows that confirmation of $H_s \approx 0.^{55}$

Likelihoodism helps specify the information that is lacking in evidential assessment of H_s. On this account, evidence 'favors' one hypothesis over an incompatible alternative just in case their likelihoods differ. The likelihood framework only applies to cases for which there are empirical grounds for assigning likelihoods to incompatible alternative hypotheses with respect to a single body of evidence. When this information is available, likelihoodism and Bayesianism coincide. The likelihood ratio for H_p is, by assumption, very favorable: e_g strongly favors H_p over its negation. For H_s, however, further information is needed before a likelihood value can be assigned to either alternative. The relevant facts concern the distribution of G-values in P because these determine what data can be predicted given H_s. If the distribution is tightly clustered around the mean, then P is very nearly homogeneous with respect to G. If some cells in P have g_1, then all (or nearly all) do, and similarly if S has g_1. In this case, evidential support for H_p and H_s is nearly the same. However, if the distribution of G-values is otherwise, such that P is heterogeneous with respect to G, the degree of support may be very different. The likelihoodist verdict hinges on information that stem cell experiments do not provide, at least for many traits. Without this information, evidential support for H_s cannot even be assessed, while e_g's strong support for H_p is unequivocal.

Mayo's error statistical account (1996) yields the same result, from a different perspective. Rather than a central theorem defining the evidence–hypothesis relation, Mayo's view is that evidence cannot be evaluated apart from experimental methods. To have good evidence for a hypothesis H, according to the error statistical view, is to have a good, or 'severe,' test for H on a set of data e. Passing a severe test requires that e fits the data predicted by H, and that the probability is high that, were H false, the experiment would not yield data that fit H, as well as e. That is, the method used to produce e must rule out relevant sources of error: alternatives to H that could produce data similar to that predicted by H. In the case at hand, the alternatives are null hypotheses $\neg H_p$ (no cells in P have g_1) and $\neg H_s$ (cell S does not have g_1). Again, H_p fares well: e_g is a good fit to H_p's prediction, while it is very improbable that this data would be observed if $\neg H_p$ were true. But, single cells are below the level of resolution for most stem cell experiments. Indeed, the very procedure that enhances severity of the test for H_p, pooling cells to increase the disparity between treatments and controls, obliterates any distinction among pooled cells with respect to G. If S's having g_1 indicates nothing about whether other cells in P have g_1, then data predicted by H_s does not fit e_g; indeed, there is no prediction and so, trivially, $\neg H_s$ is not ruled out. Without information about the variance of G-values in P, severity for H_s cannot be assessed.

3.3.3 Evidential gap

All three philosophical accounts yield the same verdict: for stem cell experiments to yield good evidence for H_s, more information is needed about the distribution of G-values in P. If cells in P are homogeneous with respect to G, then the evidential gap between H_p and H_s is minimal. But, if G-values in P vary widely, then good evidence for H_p is no evidence at all for H_s. So, evidential support for hypotheses about individual stem cells hinges on the assumption that measured cell populations are homogeneous with respect to characters of interest.[56] The greater the variation in character values for g among individual cells in P, the weaker the evidence for H_s. Support for H_s by stem cell experiments thus depends on an additional premise:

P1 Stem cell experiment M isolates cell population P from organismal source OS.
P2 M reliably measures character G in P, yielding data e_g.
P3 e_g strongly supports H_p (Some cells in P have g_1).
P4 If H_p is true, then (probably) all cells in P have g_1.

C H_s is true (cell S has g_1).

P4 is a homogeneity assumption: if some cells in P have g_1, then (nearly) all do. If true, this premise licenses inference from e_g to H_s, securing evidential support for single-cell hypotheses. However, there are reasons to doubt that P4 is true for stem cell populations in general. To be sure, *some* characters are likely to be invariant. Because measured cells share a common ancestor (ultimately, the source organism zygote), they are genetically homogeneous apart from mutations acquired during division. Environmental variation is reduced by maintaining cells under identical conditions; samples are from the same culture plate, organ colony, or animal body. Certain techniques also increase homogeneity of measured populations. For example, as noted above, microscopic observations are used to select colonies with a 'uniform' appearance in culture, while FACS subdivides a heterogeneous population into homogeneous subsets – relative to a set of characters.

But for stem cell signature traits – the crux of experimental results – there is a strong case against P4. Cell culture induces genetic changes and imposes strong selection on cell characters, particularly those involved in cell division and colony formation. Moreover, the characters of interest for stem cell biology are not DNA sequences per se, but expressed genes and proteins, morphological and functional characters. These characters are notoriously variable – even for genetically-identical cells; indeed, stem cell experiments exploit this very plasticity. Because even standardized artificial cultures are not perfectly uniform, subtle variations within cell populations cannot be eliminated. Cell–cell interactions are another source of diversity. In multicellular organisms, a cell's immediate environment is largely composed of other cells, which interact physically, chemically, and (in some cases) electrically. Signals among neighboring cells play a major role in development and can lead to 'microscale' variation, even between adjacent offspring of one parent cell. The concept of a cell niche, a microenvironment composed mainly of other cells that controls the developmental pathway taken by a given cell, was originally proposed to explain heterogeneity among tightly-packed hematopoietic cells in bone marrow (Schofield 1973). Artificial environments do not eliminate this variation. ESC cultures, for example, exhibit 'spontaneous differentiation' when cells become crowded together.

To sum up: stem cell populations, whether extracted directly from an organism or allowed to interact in culture, are known to vary in many respects. So homogeneity cannot be assumed. This undermines grounds for P4 and thereby evidential support from stem cell experiments for hypotheses about individual stem cells.

3.3.4 Scientific response

One response to this situation is to reduce the evidential gap between experimental measurements and stem cell hypotheses. The most obvious way to do this is to increase experimenters' access to individual cells. Support for H_s is then straightforward:

P1   Stem cell experiment M isolates cell S from organismal source OS.
P2   M reliably measures character G for S, yielding data e_g.
P3 e_g strongly supports H_s (cell S in P has g_1).

C H_s is true (cell S has g_1).

To a great extent the 'single-cell standard' defines experimental progress in stem cell biology. A number of sources attest to this.

Current 'gold standards' for stem cell experiments emphasize the importance of isolating and measuring single cells. For tissue-specific stem cells, the gold standard is a single-cell transplant leading to long-term reconstitution of an animal's tissue or organ. In the other branch, an ideal pluripotent stem cell line behaves as a single cell, exhibiting the same traits in the same culture environment, so self-renewal or differentiation capacities can be realized on demand according to researchers' specifications.[57] Post-genomic and imaging technologies that enhance our ability to isolate or track single cells are quickly adopted by stem cell biologists and reported as advances in the field.[58] The single-cell standard dates back at least to World War II-era experiments with cultured cells and transplantable tumors in inbred mice. The first method for measuring stem cells was announced as "a direct method of assay for [mouse bone marrow] cells with a single-cell technique" (Till and McCulloch 1961, 213).

Moreover, the 'single-cell standard' is frequently invoked in statements about future progress. For example:

Comparisons [of stem cells from different organismal sources] require isolating pure cells, with minimal culturing, at a defined state of development in sufficient quantities such that comparative methods can be performed with sufficient rigor to overcome the variability inherent in the comparison techniques itself.

(Cai et al. 2004, 586)

...concerns have been raised up [sic] regarding cell purity and introduction of artifacts to the follow-up analysis [of stem cells]. This implies a fundamental need to study stem cells at the single-cell level... Progress... especially in rare and heterogeneous stem cell populations, is dependent on advances in single-cell assays.

(Wang and Audet 2009, 337)

Perhaps the only way to truly understand the reprogramming process [for iPSC] will be to extend the recent studies that combined single-cell analysis with fine temporal resolution.

(Plath and Lowry 2011, 263)

All this suggests that new technologies are evaluated by stem cell biologists with respect to the single-cell standard, such that those which improve our 'prospective grasp' of stem cells are deemed successful. Publication and citation patterns seem consistent with this proposal, with new experimental methods predominating over new hypotheses or models. It is important to note that this experimental emphasis undercuts the traditional dichotomy of science and technology, rather than reinforcing it. Stem cell biologists' emphasis on methods and technology is grounded in evidential considerations. Technical innovations that improve our access to single stem cells allow experiments to better support hypotheses about stem cell signatures.

3.4 Population-level models

A simpler response to the evidential problem above is to confine stem cell hypotheses to claims about cell populations. Models of stem cell traits and capacities need not describe every individual cell; indeed, such a level of detail might undercut the explanatory or predictive power of stem cell hypotheses. Representational assumptions R2 and R3 relate the abstract stem cell model to cell populations and types, respectively, so there is no fundamental conceptual barrier to hypotheses about these entities. Two kinds of model, stochastic and compartmental, yield hypotheses about stem cell populations.[59] However, close examination shows that this simple 'avoidance' response will not do. Evidential support for these R2/R3 models depends on hypotheses about the traits of single stem cells. So, population-level stem cell models do not avoid the evidential problem posed by cell heterogeneity. A seminal case – the first population-level stem cell model and its empirical basis – demonstrates the point.

3.4.1 Stochastic model

Any population of cells experiences some number n divisions over a period of time τ, such that the population grows, diminishes, or remains constant in size. Any dividing cell in the population has a certain probability of undergoing each of three kinds of division (Figure 3.4): both offspring like the parent (p), one offspring like the parent (r), or no offspring like the parent (q), where $p + r + q = 1$. In this basic model, relations among p, r, and q values entail general predictions about cell population size (growth, decrease, or 'steady-state'), and equations that predict mean and standard deviation in population size, probability of stem cell extinction, and features of steady-state populations are derived.[60] In these equations, which comprise the 'p-r-q model,' p is the fundamental parameter (Till et al. 1964, Vogel et al. 1969). Testable

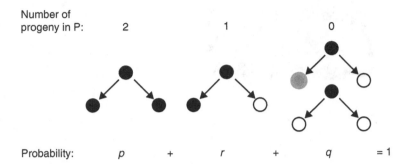

Number of
progeny in P: 2   1   0

Probability: *p*   +   *r*   +   *q*   = 1

Figure 3.4 Simple stochastic stem cell model, representing probabilities of cell division events

predictions require that its value be estimated. This is done by estimating the coefficient of variation for stem cell number in populations of the same age produced by division from a single founding stem cell. The data required for such an estimate are numbers of stem cells in replicate colonies, each originating from a single stem cell. Though not generally available for developing cells, such data are provided by 'clonal' stem cell assays. The original assay of this kind is the spleen colony assay (Till and McCulloch 1961).

3.4.2 Spleen colony assay

The spleen colony assay was the first method for quantitatively measuring stem cells and furnished the first experimental demonstration of stem cell capacities. Like many scientific advances, it began as a serendipitous byproduct, in this case of post-World War II studies of the effects of radiation on mammalian cells and their reversal by bone marrow transplantation (Brown et al. 2006, Fagan 2010, Kraft 2009). About a week after bone marrow injection, lumpy nodules containing ~1 million cells appear on spleens of surviving mice (Figure 3.5a). In 1961, James Till and Ernest McCulloch of Toronto's Ontario Cancer Institute harnessed this phenomenon in a quantitative assay, which fits the general pattern of stem cell experiments (Figure 3.5b). The organismal source is adult mouse bone marrow. Cells from this source are mixed to eliminate clumps, counted, and a known number injected into lethally-irradiated mice.[61] After 10–11 days, surviving mice are killed and the second measurements are made: number of spleen colonies, types of blood cell within a colony, and proportion of cells of each type.

The spleen colony experiment yields pairs of data-points correlating the two measurements: number of cells injected and number of spleen

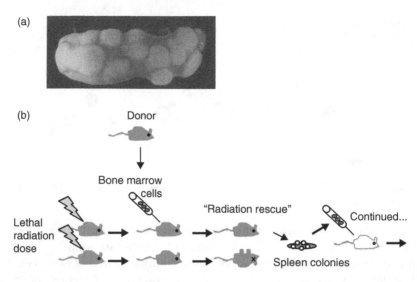

Figure 3.5   Spleen colony assay. (a) colonies from mice given lethal doses of radiation followed by injection of 10^5–10^6 bone marrow cells. Reprinted from Till et al. (1970), with permission from AAAS. (b) Design of mouse irradiation and spleen assay experiments

colonies. The relationship is approximately linear: ~1 spleen colony per 10^4 injected bone marrow cells. Because many cells are injected, evidence that spleen colonies have a single founder cell is perforce indirect.[62] The clearer result was that 'spleen colony-forming units' (whether one cell or more) divide to produce mature blood cells of different types, easily distinguished when colonies were dissected. Colonies also contained less-differentiated cells, which produced new colonies upon injection into irradiated mice. This result demonstrated that 'primary' colonies include colony-forming cells. However, numbers of these cells within a colony and the number of colonies formed per injected cell were highly variable, even with other conditions held constant.[63] This is exactly the experimental arrangement required to test the p-r-q model – unsurprising, as the model was developed with precisely this case in mind.

3.4.3   Evidence for population-level models

The p-r-q model predicts features of cell population kinetics, which fit fairly well with experimental data. But the hypothesis thereby supported is *not* that spleen colony-forming cells are stem cells. Rather, it is the hypothesis that stem cell population size is regulated so as to yield predictable population-level results from randomly-distributed single-cell capacities. Testing this hypothesis requires identifying stem cell populations. The p-r-q model predicts features of cell population kinetics for spleen colonies, *given* the assumption that colony-forming units are stem cells. All these predictions hinge on estimation of the fundamental parameter *p*: the probability that a stem cell undergoes self-renewal. This parameter is estimated from the pattern of variation in a set of replicate colonies, which is assumed to be initiated by a single 'stem element.' Spleen colony-forming cells satisfy the conditions for a stem element. But, in order for experiments to be *replicates*, all the stem elements for the set of colonies must be assigned the same probability values for *p* and $(1-p)$-ω; that is, they are assumed to have the same capacities for self-renewal and differentiation. Experimental test of the p-r-q model depends on the assumption that the cell population measured is homogeneous with respect to these characters. In this sense, evidential support for the stochastic p-r-q model depends on a hypothesis H_s, in which *G* refers to stem cell capacities.

This evidential constraint is not peculiar to the p-r-q model.[64] The above argument applies to any population-level stem cell model which assigns probability values to single stem cells. Evidence for such a model presupposes a homogeneous stem cell population, the 'starting pool' for experimental replicates. R2-models in general, then, do not avoid the evidential gap discussed above (see §2.4.2). R3-models, whose objects are interpreted as reproductively-related cell types, do no better. These models classify cells into discretely bounded types, each defined by a set of character values C_i. Cell differentiation is represented as transfer from one type, or 'compartment,' to another, while self-renewal is division within a compartment (Figure 3.6). These 'compartment models' represent stem, progenitor, and mature differentiated cells as a system of linked compartments exhibiting population dynamics, which can be represented by a system of differential equations. Solutions to these equations entail precise predictions about changes in cell population size for different compartments. But, as with stochastic models, all these predictions depend on the assumption of homogeneous cell compartments.

Though stem cell capacities are defined as population averages in R3-models, the character values that define each cell type are not. Cells of a given type are homogeneous with respect to C_i. So to make predictions, R3-models must presuppose H_s for $G = C_i$, where C_i includes stem cell traits. As with R2-models, these single-cell hypotheses are not part of the model as such, but are a condition of deriving predictions that can

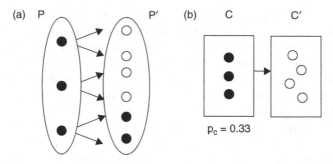

Figure 3.6   Cell compartment concept. (a) Cell populations related by division. P is the parent generation; P' the offspring generation. (b) Cell compartments related as successive developmental stages, C and C'. C is decreasing, as the proportion of offspring that remain in C (self-renewal) is <0.5

be tested by stem cell experiments. In general, then, population-level stem cell models do not avoid the evidential problem posed by cell heterogeneity. Homogeneous stem cell populations are a prerequisite for their evidential support. The single-cell standard is thus the best evidential strategy for stem cell models using any of the representational assumptions R1–R3.

3.5   The 'stem cell uncertainty principle'

Guided by this standard, technical innovations since the 1950s have increased our ability to manipulate and measure character values of single mammalian cells, including size, surface molecules, and gene expression. In this sense, stem cell biology has progressed beyond the original exemplary methods described above. But another, deeper, evidential problem remains: the gap between experimental data and hypotheses about stem cell capacities. As discussed in §3.1, stem cell experiments involve two sets of measurements, both of which provide data about characters of single cells. *But no single cell persists through both sets of measurements.* Cells reproduce by division, so descendants and ancestors cannot coexist. The second set of measurements is of cells descended from those measured in the first. Self-renewal and differentiation potential are measured after realization of these capacities in controlled environments: the second set of measurements.[65] A single stem cell, therefore, can be identified only retrospectively. The single-cell standard, rigorously applied, means stem cell researchers literally don't know what they've got 'til it's gone.

Technical innovations cannot remedy this evidential situation, as the latter follows from the basic structure of stem cell experiments and the stem cell concept itself. There are actually three distinct evidential problems here. First, self-renewal and differentiation potential cannot both be measured for a single cell, even if single-cell technologies are available. To determine a cell's differentiation potential, that cell is placed in an environment conducive to differentiation and its descendants measured. To determine a cell's self-renewal ability, the cell is placed in an environment that is conducive to cell division *without* differentiation and its descendants measured. It is not possible to perform both experiments on a single cell. Because stem cells are defined as having both capacities, they cannot be identified at the single-cell level. A second problem is that capacity for self-renewal cannot be decisively established for any stem cell. An offspring cell with the same capacities as a stem cell parent has the same potential for differentiation and for self-renewal. Even if both could be measured for a single cell (which they cannot), it is the offspring of the offspring cell that indicates the latter's capacities. The relevant data are always one generation in the future. Experimental proof that a single cell is capable of self-renewal is infinitely deferred. Differentiation potential presents a third evidential problem. In any experiment, this potential is realized in a range of (highly artificial) environments. But these data cannot tell us what a cell's descendants would be like in a different range of environments – in particular, physiological contexts. There is, inevitably, an evidential gap between the capacities of a cell un-manipulated by experiment and realization of capacities in specific, highly artificial, contexts. For all three reasons, claims that any single cell is a stem cell are inevitably uncertain.

This uncertainty admits diverse, even arbitrary, operational criteria for self-renewal and underpins perennial debate over the extent of differentiation potential in stem cells from adult organisms. These evidential limitations of stem cell experiments have been likened to the Heisenberg uncertainty principle, which states that a particle's mass and velocity cannot be measured simultaneously. In physics, the procedure used to determine the value of one alters the value of the other. The analogy suggests that measurement itself is the problem, that "...we cannot determine both the function of a cell and its functional potential...[because] our determination of a cell's function at a given point in time interferes with an accurate determination of its developmental potential" (Nadir 2006, 489), and we cannot rule out the possibility that "the investigator might be forcing the stem-cell phenotype on the population being studied" (Zipori 2004, 876). However, in stem cell biology, the problem is not measurement of cells per se, but their transfer to different environmental contexts. Stem cell capacities are realized and measured in cells descended from 'candidate' stem cells in different environments and with different traits. Potten and Loeffler (1990, 1009) articulate the issues incisively:

> The main attributes of stem cells relate to their potential in the future. These can only effectively be studied by placing the cell, or cells, in a situation where they have the opportunity to express their potential. Here we find ourselves in a circular situation; in order to answer the question whether a cell is a stem cell we have to alter its circumstances and in so doing inevitably lose the original cell and in addition we may see only a limited spectrum of responses... Therefore it might be an impossible task to determine the status of a single stem cell without changing it. Instead one would have to be satisfied with making probability statements based on measurements of populations.

Stem cell experiments, no matter how technically advanced at tracking and measuring single cells, cannot resolve stem cell capacities at the single-cell level. This is because we cannot directly measure a single cell's capacities for self-renewal or differentiation, separately or together. So, just as evidence for population-level models of stem cells depends on single-cell measurements, evidence for single-stem-cell models depends on measurement of cell populations. To measure both self-renewal and differentiation potential for a single cell, and to elicit the full range of a cell's potential, multiple 'copies' of that cell are needed – a homogeneous cell population of *candidate* stem cells. The evidential gap between stem cell capacities and measured character values shows up clearly in argument form. Let H_{sc} be the hypothesis that cell S is a stem cell with capacities C_{sc} and H_{SD} the hypothesis that cells in population P self-renew for time interval τ and differentiate into cell types $T\iota$. Experimental support for H_{sc} is as follows:

P1   Stem cell experiment M isolates cell population P from organismal source OS.
P2   M reliably measures stem cell capacities in P, yielding data e_g.
P3 e_g strongly supports H_{SD}.
P4   If H_{SD} is true, then (probably) all cells in P have capacities C_{sc}.

C H_{sc} is true (S is a stem cell with capacities C_{sc}).

If P is a homogeneous population of candidate stem cells, then there are good grounds for P4. Conversely, heterogeneity in P weakens the evidential tie between H_{SD} and H_{sc}.

Experimental controls and single-cell methods can bolster the grounds for P4. A rigorous 'clonal' experiment M begins with a single cell in a controlled environment, with all relevant signals that could impact the cell taken into account. If all other cell reproduction in this environment is blocked or products of the founding cell can be distinguished from all other cells, then e_{sc} reflects the reproductive output of a single starting cell and no others. Measured stem cell capacities can then be unambiguously attributed to that cell *in that environment*. Technologies that track a single cell's reproductive output over time, such as real-time imaging and cytocinematography, combined with techniques that measure character values of single cells, can yield data of this sort. In this way, technical innovations guided by the single-cell standard can bolster P4. But they do so only relative to the environment in which stem cell capacities are realized. More general results are obtained from replicate clonal experiments using a range of environments. If replicates behave consistently, this shows that the same environment tends to elicit self-renewal of the same duration and/or differentiation into the same cell types, while different environments reliably yield different results. Experimental results that are robust in these ways indicate that the cell population from which replicates are drawn is homogeneous with respect to stem cell capacities. Of course, populations homogeneous with respect to *one* set of cell traits need not be homogeneous with respect to *all*. But, sorting cells into populations homogeneous for many measurable traits is the best way to proceed, given that we cannot identify stem cells in advance.

So, the 'stem cell uncertainty principle' does not undercut all support for hypotheses about single stem cells. But, as the possibility of heterogeneity in stem cell capacities cannot be completely ruled out, stem cell hypotheses can never be fully and decisively established. Experimental support for these hypotheses depends on the assumption that the starting population is homogeneous. For some characters, notably those included in the first set of measurements, homogeneity is imposed during the experiment. Stem cell experiments can therefore provide good evidence for stem cell hypotheses at the single-cell level, but only relative to the set of characters used to specify a homogeneous subpopulation in the first set of measurements. As new cell traits are discovered and made accessible to measurement, the assumption of homogeneity must be continually reassessed and revised. Moreover, experimental results only indicate stem cell capacities relative to the set of environments used in the second set of measurements. All hypotheses about single stem cells are therefore doubly relative: to the set of characters *and* environments used in the experiment. Such hypotheses are necessarily provisional, becoming obsolete when new characters and environments are introduced.

The field of stem cell research is a patchwork of experiments, each supporting different specific hypotheses relative to the method used. The question then arises: Can stem cell experiments support more general hypotheses? If not, the field is inherently disunified. Later chapters argue that general and robust claims about stem cells emerge at the community level by combining data from many diverse experiments. Chapter 4 examines another response: conceptual revision. As the evidential limitations on stem cell experiments are rooted in the stem cell concept, why not jettison or modify it? A number of stem cell biologists have made proposals along these lines, concentrating on the notion of 'stemness.'

3.6   Conclusions

This chapter examines two evidential challenges for stem cell experiments: the inferential gap between single cells and cell populations, and between stem cell capacities and the cells in which they are realized. The gap between experimental data about cell populations and hypotheses about single cells is bridged by a 'homogeneity assumption' (P4), to the effect that cell populations measured are homogeneous with respect to traits of interest. The homogeneity assumption is difficult to justify and may be false for many candidate stem cell populations. Restricting stem cell hypotheses to cell populations or types does not avoid the problem, as experimental support for these hypotheses depends on claims about characters of single cells. However, technical innovations that increase experimenters' ability to measure and track single cells can bring about a situation in which experiments can provide strong evidence for hypotheses about stem cells. 'Single-cell' technologies are thus an important form of progress in stem cell biology, strengthening inferences from experimental data to conclusions about cells and their traits. These results account for the prevalence of the 'single cell standard' in stem cell biology, which has held sway for over four decades. They also explain why technical innovations are the main form of progress in stem cell biology, and prioritized over population-level models and abstract modeling. Importantly, these considerations undercut the traditional science/technology dichotomy. In stem cell biology, technical innovations have clear and vital evidential significance.

The second problem, formulated as 'the stem cell uncertainty principle,' arises from the solution to the first. The 'principle' states that stem cell capacities, realized only in descendants, are impossible to measure directly; a single stem cell can be identified only retrospectively. It follows that self-renewal and differentiation cannot both be measured for the same cell, nor can alternative developmental pathways be measured for a single cell. Stem cell experiments thus appear to face an evidential dilemma: stem cell hypotheses and models depend for evidential support on hypotheses about single stem cells, but, due to unavoidable uncertainty, hypotheses about single stem cells cannot be decisively established by experiment. Technical innovation cannot ameliorate the difficulty, which is grounded in the stem cell concept. But, the evidential situation is not so dire as it might seem. Stem cell capacities can only be reliably inferred relative to an experimental context, not absolutely. An experimental context includes the traits relative to which stem cells are homogeneous and the environments in which stem cell capacities are measured. The best way forward is to recognize these evidential limitations, and treat all stem cell hypotheses supported by experiment as provisional and context-relative.

Together, these limitations raise concerns about lack of generality and robustness for well-supported stem cell hypotheses. Given pure populations of identical stem cells, we learn what happens to a stem cell of a given type in a range of environments. However, there are no grounds for generalizing beyond the range of environments used in experiments. Inferences about stem cell capacities are relative to an experimental method, with a specific organismal source, characters measured, and manipulations of cell environment. The main inferential challenge in stem cell biology is not generalization from a sample or exemplar, but coordination of different experimental results. Later chapters examine this problem in detail.

4
A State of Uncertainty: Stemness and the Roles of Theory

4.1 The stemness alternative

The stem cell uncertainty principle shows that evidential significance of stem cell experiments hinges on details of the method used (Chapter 3). Insofar as they are empirically-confirmed, hypotheses about single stem cells are relative to organismal source, characters measured, and manipulations performed within a temporal duration of interest. Evidence for population-level models also depends on such hypotheses. These evidential constraints create serious challenges for a research program aimed at prospectively identifying stem cells on the basis of 'signature' traits. To overcome these limitations, a number of stem cell researchers propose a conceptual revision: replace the notion of 'the stem cell' as cellular entity with that of a cell *state*, 'stemness.' The idea of a cell state is usefully contrasted with that of a cell *type*.[66] A cell type is defined by a set of character values, such that a cell is classified as a particular type in virtue of having all (or most) of those character values. Characters in the set vary, ranging from structural to functional, morphological to biochemical, features of the cell as a whole, the cell surface, molecules, and internal structures. All, however, are conceived as characters *of the cell*. In contrast, a state is a functional role taken by a cell, within some larger process.

In this chapter, I set out the stemness alternative (§4.1), relate it to the abstract model of Chapter 2 (§4.2), and examine the stemness critique of other views of development (§4.3). I then offer a critical response, building on work by Cartwright and colleagues on trade-offs among the roles of theory in physics (§4.4). Extending these ideas to the stem cell case, I argue that the stemness alternative is designed to fill multiple roles that are in tension with one another: a general definition (§4.5), a source of empirically accurate predictions (§4.6), and an explanatory sketch (§4.7). Owing to conflicts among these roles and aspects of the stem cell concept discussed in previous chapters, the stemness alternative cannot fulfill its tasks. This accounts for its lack of uptake by the wider stem cell community. However, a more modest stemness alternative, focused on molecular genetic explanations of cell development, is a valuable contribution that deserves wider attention.

I begin by setting out the stemness alternative in more detail. Its most developed explication and defense is by Dov Zipori (2004, 2005, 2009). His view consists of four main theses:[67]

(i) "...cells have several states of existence, such as proliferation, differentiation, and the 'stem state.'" (2004, 877);

(ii) The stem state is "characterized by instability" (2009, 204) and "defined by the highest degree of plasticity of a cell, within the repertoire of cell types present in the organism" (2005, 719);

(iii) "Stemness can be defined as a state of having all options open, whereas differentiation is a process wherein the options... that are harboured by the cell are diminished" (2004, 875);[68]

(iv) "Stemness might be a transient and reversible trait that almost any cell can assume given the correct trigger" (2004, 876).

On Zipori's view, the stem state is defined not in terms of cell character values, but possible outcomes of a process – differentiation. More specifically, stemness is a state from which many other cell states can be reached via differentiation. Zipori describes differentiation as both state (i) and process (iii). The two ideas can be reconciled by conceiving the developmental process by which a cell acquires specialized traits as a sequence of states occupied by the cell. Each developmental process, or pathway, terminates in a mature, specialized cell. In graphical terms, the stem state is a node from which many such pathways emanate.

The defining attribute of stemness is not a set of cell character values, but accessibility of many differentiation states with distinct developmental outcomes. Zipori refers to this structural relation of stemness to differentiation as "plasticity," which he conceives as "a hallmark of the stem state:"

Stemness seems to be a standby state in the cell life cycle: being in this state may entail a lack of determination in gene expression, and availability of a great number of differentiation options, i.e. plasticity. The latter is referred to herein as a state in which the cell harbors a potential to give rise to a multitude of lineages.

(2009, 169)

The stemness alternative thus substitutes for two complementary reproductive capacities of cells, two complementary cell states: stemness and differentiation. Self-renewal drops out as an essential feature. In differentiation, a cell's developmental options are progressively restricted as it acquires more specialized traits. Stemness is characterized by the absence of specialized traits and a wide range of developmental options. Cell development is conceived as a series of transitions between complementary states; one maximizing plasticity, the other reducing it.

Zipori's 'plasticity' definition entails several important consequences. First, the stem state is *transient*, a "pause" in the process of cell differentiation. In leaving it, a cell "chooses" one of many possible developmental pathways. This choice is, of course, not a conscious decision, but determined by a combination of internal and external factors, or "signals," impinging on the cell. Second, such choices are, in principle, *reversible*. Just as cells can be induced by signals to leave a stem state, so can they be induced to enter such a state. More generally, the 'topology' of developmental pathways is determined by the probability that a cell transitions from an initial state to other states. Cell development is conceived as a continuum, different positions on which correspond to different functional roles that cells can occupy. Cells occupy functional roles in virtue of being in particular states. Every cell in a multicellular organism has some probability of assuming each state, including the stem state. In general, this probability declines as cells differentiate, but there is no sharp boundary or in principle barrier to stem cell function. Stem cells are just all and only those cells occupying states with the option of many different developmental pathways. So a third consequence of Zipori's stemness account is that stem cells are *interchangeable*, not relative to organismal sources or experimental methods.

Though not widely-discussed among stem cell biologists (at least in the United States), Zipori's view is neither unique nor idiosyncratic. Similar accounts include "the evolving stem cell concept" (Blau et al. 2001), "chiaroscuro stem cell model" (Quesenberry et al. 2002), and the "arrested development" proposal (Mikkers and Frisén 2005). Table 4.1 summarizes their main similarities and differences. According to the evolving concept, to be a stem cell is to occupy a specific functional role: giving rise to more differentiated cells and regenerating adult tissues.

Table 4.1 Comparison of different stemness accounts

	Stem state	Stem cell capacities	Stem cell signature	Explanation	Developmental process
Zipori	Functional role	DF only	No	Molecular	Reversible
Blau et al.	Functional role	DF only	No	ND	Reversible
Quesenberry et al.	True capacities revealed	SR and DF	Yes	Molecular	Reversible
Mikkers and Friesen	Functional role	DF only	Yes	Molecular	Irreversible

DF: differentiation; ND: not discussed; SR: self-renewing.

This function, Blau and colleagues argue, "can be induced in many distinct types of cells, even differentiated cells" (2001, 829). The chiaroscuro model, in contrast, assumes that stem cells are "cellular entities" with stable capacities for self-renewal and differentiation. "Chiaroscuro" refers to techniques of representing light and shadow in painting. As representations of objects contrast when depicted in light or shadow, by analogy, stem cell capacities may be revealed or hidden. In the stem state, they are revealed – but the situation may change. A single cell can be a stem cell with wide differentiation potential at one time, a progenitor restricted to one cell fate at another and a progenitor committed to a different cell lineage at yet another. The methodological implications are the same as for stemness: "in a nonsynchronized population of cells [that is, nearly all cell populations], at any one point in time, a number of true stem cells would not be detectable" (Quesenberry et al. 2002, 4270). Finally, Mikkers and Frisén (2005) define a stem cell as a cell that resists (for a time) the flow of developmental progress. Apart from the assumption that cell development is linear and irreversible, their "arrested development" view is much like Zipori's, conceiving stemness and differentiation as complementary functional states, which developing cells can occupy.

4.2 Contrasting models

The above accounts converge on a "radiating" model of cell development, which graphically represents the stemness alternative (Figure 4.1). On this view, all and only stem cells, by definition, have wide differentiation potential; i.e., plasticity. But to what exactly is this radiating model an *alternative*? A minimal structure, it closely resembles the abstract stem cell model of Chapter 2, which defines a stem cell as the apex of a hierarchy organized by reproductive relations of self-renewal and

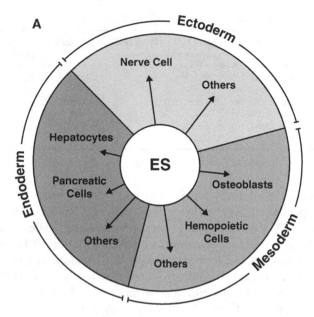

Figure 4.1 'Radiating' model of the stem state for pluripotent ESC. After Zipori (2005) with permission from John Wiley and Sons

differentiation (§2.6, Figure 2.7). Yet, structurally, the abstract model contrasts with the stemness account in two ways. First, self-renewal is essential in the former but not the latter.[69] Second, the abstract model defines stem cells as having maximal differentiation potential relative to a lineage of interest, allowing for a wide range of "potencies," while the stemness account defines stem cells as pluripotent, the source of a "burst" of radiating developmental pathways.[70] That the stemness account conceives development as continuous, with no sharp distinction between stem, progenitor, and fully differentiated cells, might seem a third contrast, especially given the hierarchical structure of Figure 2.7. But, in fact, the abstract model is compatible with a 'continuum' view, particularly if the temporal duration of interest is short.[71] So stemness departs from the abstract model only in requiring pluripotency, but not self-renewal.

However, outright conflict between the abstract model and stemness account is averted by their different purposes. The purpose of the abstract model is to represent the common core of diverse working definitions of "stem cell," not to guide and assess experimental methods. In contrast, the purpose of the stemness model is to offer a general definition of

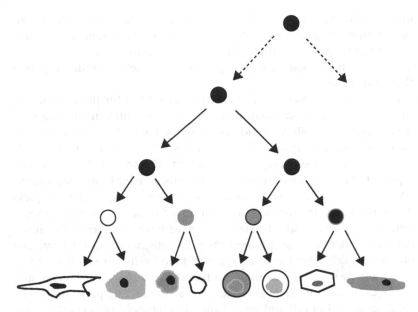

Figure 4.2 The traditional model of development, organized hierarchically by unidirectional lineage relationships

'stem cell' that directly engages experimental methods and their results. Though both aspire to generality, their different relations to experiment diffuse any direct conflict between the definitions. But if the abstract model is supplemented with certain representational assumptions, conflict with the stemness account does ensue. These assumptions are: (a) objects in the model represent developing cells, singly or in populations; and (b) relations in the model describe exceptionless generalizations, laws of development. So interpreted, the abstract model amounts to a general theory of cell development as a unidirectional sequence of hierarchical stages that progressively partition an organism into discrete tissues and organs (Figure 4.2). I will refer to this as the traditional model of development, to distinguish it from the abstract model per se.

The stemness account clearly conflicts with the traditional model as a theory of development, rejecting unidirectional pathways, fixed cell hierarchies, and exceptionless generalizations describing the sequence of developmental stages. This conflict also has implications for experiment. A natural way to apply the traditional model experimentally is to extend classificatory strategies for mature cells 'backwards' along the developmental pathways to less-specialized ancestors. Such methods implicitly assume that stem cells, like fully-differentiated cells, can be identified by distinctive set of morphological and functional traits: a universal stem cell signature. The stemness account rejects the idea of a universal stem cell signature, along with the model of cell development that supports it.

The conflict of models is even sharper given the further assumption that cellular and organismal development are 'collinear' (Figure 4.3). According to this collinear model, developmental potential of a stem cell inexorably declines as organismal development advances. So stem cells derived from early embryos are pluripotent, able to give rise to any cell type of the adult organism, while stem cells derived from adult organs and tissues are multi-, oligo-, or unipotent depending on the complexity of the tissue in question. The 'ultimate stem cell' is a totipotent, fertilized egg. Tissue-specific stem cells *must* have more restricted differentiation potential than pluripotent stem cells, with minimal overlap: blood stem cells give rise to all types of blood cells but not neurons or bone; muscle stem cells to muscle cells, but not kidney or pancreas; and so on. Collinear and stemness models make conflicting predictions. The collinear model of cell and organismal development predicts that stem cells in adult organisms have tissue-specific developmental potential, while stem cells derived from early embryos (or engineered to resemble them) are pluripotent.[72] The stemness model, in contrast, predicts that developmental potential of adult stem cells exceeds their typical outcome within a local tissue. More stem cells are pluripotent, on this view, than our experiments detect. So experimental data can, in principle, adjudicate between these two conflicting models of cell development. However, in practice, experimental evidence has not been decisive, for reasons explored below (see §4.6). To adjudicate between these conflicting and contrasting models of development, we need to look more closely at the vexed "state vs entity" debate.

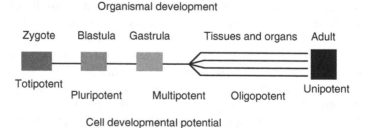

Figure 4.3   Collinear model of cell and organismal development

4.3   Evaluating models

Zipori and other proponents of stemness reject the "entity theory" that stem cells are cellular entities. Some have taken this position to be a form of anti-realism about stem cells.[73] But this is a misunderstanding. Zipori does not deny that there are cells in the stem state, which function as stem cells. What he rejects is the thesis that there is a unique set of character values had by all and only stem cells: a "stem cell signature." As discussed in previous chapters, the prevailing research program in stem cell biology presupposes that there is such a signature by which a homogeneous stem cell population could be prospectively isolated. So the stemness alternative poses a serious challenge. If there is no "discrete cellular entity" with stem cell capacities that can be prospectively isolated on the basis of cell characters, then experiments that aim to identify stem cells in this way are doomed to fail.[74] This is not to be confused with the claim that there are no stem cells. What proponents of stemness deny is that stem cells comprise a discrete population with stable character values, which can be prospectively isolated and put to use. Their arguments are about experimental methods, as well as concepts. In stem cell biology, as previous chapters show, the two are intertwined.

Zipori, whose arguments for stemness are the most developed, offers three objections to the idea of a universal stem cell signature. The first is that it is based on a flawed analogy between stem and mature cells. Fully-differentiated cells have characteristic 'molecular markers:' genes expressed in one cell type but not others. If stem cells are another type of cell, alongside neurons, lymphocytes, and so on, then they too must have characteristic molecular markers. But markers of mature cells are associated with functions that stem cells lack, so we should not expect stem cells to exhibit molecular markers. Stem cells should rather be characterized by the *absence* of molecular markers for mature cells. Second, self-renewal is not a distinctive capacity of all and only stem cells; it is not "stem-cell-specific" (Zipori 2004, 874). So there can be no universal molecular signature for stem cell self-renewal. Third, stem cell capacities are not "cell autonomous," but imposed on stem cells by their immediate external environment, or niche. So methods that disrupt the association of stem cell and niche cannot reveal a universal stem cell signature. And if stem cells are conceived as autonomous from their environments, there can be no such signature, full-stop. The stemness account avoids all three objections, invoking a cell state defined by plasticity, occupied by cells in a niche, with no requirement for self-renewal.

These objections mingle conceptual and empirical issues, though more emphasis is placed on the former. Empirical evidence comes to the fore in Zipori's arguments about adult stem cell plasticity and recent attempts to identify a stem cell signature using new 'high-throughput' methods associated with systems biology. I examine these arguments below (§4.6). Here, it is important to note that previous chapters offer support for Zipori's conceptual claims, especially Chapters 2 and 3. Chapter 2 showed that the concept of self-renewal is deeply method-relative, differs across the two main branches of stem cell research, and cannot be unambiguously measured even in the most rigorous experimental settings. Therefore, no context-independent specification of self-renewal can be represented as an attribute of all and only stem cells. The crucial role of stem cell niches is detailed in Chapters 2 and 3. Every experimental context in which stem cells are studied includes features of cells' environments, either in artificial culture or within an organism. Indeed, manipulations in stem cell experiments are just changes to a cell's environment. Zipori's first objection is less decisive. The insight that stem cells lack the distinctive specialized features of mature cells is part of the prevailing research program. 'Negative selection' by contrast with mature cells is included in the exemplary methods for identifying stem cells (see Chapter 3). Nonetheless, in light of previous chapters, Zipori's objections merit serious consideration. Yet they have elicited little response from the wider stem cell community. Philosophy of science sheds light on this confusing situation.

4.4   Philosophical responses

Happily, there is a direct philosophical response already on offer. Leychkis et al. (2009) examine Zipori's stemness account, compare it to an alternative, and conclude that there is currently insufficient evidence to rule out or accept either. They further suggest that improved versions of state and entity theories may be compatible. These are reasonable conclusions. Yet, in another respect, Leychkis and colleagues miss the point of the stemness debate, focusing on theories rather than models and reformulating scientists' language to meet philosophical standards. So, although Leychkis et al. take pains to engage the biological facts as currently understood and carefully examine experimental methods, their analysis does not cut as deeply as it might. After critically summarizing their account, I offer another interpretation of the stemness debate.

Leychkis et al. describe "two rival approaches" to stemness as competing theories: the historically prior entity theory and the "newcomer" state theory (2009, 312–313). The former is represented by experimental reports published in 2002 by two East Coast laboratories. These reports describe genes expressed in various kinds of stem cell [ESC, HSC, and neural stem cells (NSC)] that are not expressed in cells lacking stem cell capacities. The studies presupposed that stem cell capacities have a genetic basis, and sought to identify the genes responsible for stemness. Leychkis et al. take this project to encapsulate the entity theory "that stemness is a property of a cell – namely, the property of having a certain gene or genes whose expression makes possible both the potential for self-renewal and the potential for multilineage differentiation" (2009, 313). They distinguish strong and weak versions of the entity theory. The former claims that certain genes and their expression are the best causal explanation of stem cell functions, while the latter asserts only correlation of gene expression and stem cell function. Examining the techniques used by the two groups in detail, Leychkis et al. argue that progress on the entity theory has been through "development and qualification" of the strong version to a more qualified and nuanced view, with innovative use of new technologies playing a crucial role (2009, 314–315).

Zipori's state theory comes in for harsher criticism. Leychkis and colleagues argue that it is inadequate as a descriptive definition of 'stem cell,' as a testable biological theory and as a formulation of the concept of stemness, being, respectively, too restrictive, too abstract, and too imprecisely stated. They propose a more general definition of a state as "a concrete entity's possessing a property at a particular time t" (2009, 317). Stemness is a state of the cell, which rests on the property of plasticity. This leads to some analytic awkwardness about applications of properties and predicates, to which Leychkis et al. offer technical solutions invoking higher-order predicate logic, and necessary and sufficient conditions for predicating an abstract entity (stemness) of cells (2009, 316–317).

Leychkis et al. are right to seek clarification of Zipori's account and its alternatives. But their linguistic and analytic methods are not the right tools for the job, being honed for philosophical debates about language and representation. Casting the alternatives as theories is already tendentious, as this departs from scientists' own practice. Yet this choice makes sense. Both the stemness account and its better-established rival aim to provide a general definition that is explanatory and supported by the available evidence. Traditionally, scientific representations that aspire to this role are termed theories.[75] The stemness alternative is motivated by goals of generality, explanatory power, and empirical confirmation.

If it succeeds, then the stemness account qualifies as a theory in the traditional sense.

However, this traditional conception of theory has been sharply criticized by other philosophers, notably Nancy Cartwright and her collaborators (1983, 1999; Suárez 2008). The problem is a trade-off between general explanation and accurate empirical prediction, illustrated by case studies in physics. In physics, certain fundamental equations are identified as explanatory laws, which subsume many diverse phenomena under a relatively simple formalism. These fundamental laws succeed by abstracting from the messy and complicated details of particular cases, exhibiting their common structure. But such abstract laws do not entail accurate predictions about patterns of empirical data in real situations. Many layers of corrections and additional conditions must be added before the predictions from general explanatory laws correspond to what actually happens. Often, we can do no better than an approximation or extrapolation of a very simple case. A theory that does predict the patterns of observed data is thereby empirically supported and fits the available evidence. But such a theory is usually restricted in scope: it predicts phenomena very accurately in a narrow domain, but does not subsume a wide variety of phenomena. The reason is that real situations, whether natural or artificial, are complicated. Simple, fundamental laws show themselves empirically only in special cases. So theories that correctly describe "the facts" do not provide general explanations, while, conversely, theories that explain many facts do not accurately predict patterns in empirical data.[76]

These criticisms also apply to the stemness alternative. The core problem underlying the difficulties noted by Leychkis and colleagues is that the stemness account attempts to fulfill three roles at once: a general definition of 'stem cell,' a source of predictions that fit all the available empirical evidence, and a molecular genetic explanation of stem cell function. As in physics, these tasks are best accomplished by different representations, whether termed models or theories. So the main problem with the stemness account is that it suffers from an excess of ambition. It is intended to perform three tasks – all scientifically important, but in tension with one another. Assessing its merits and defects requires teasing these different roles apart. The next sections examine each in turn.

4.5   General definition

The first task is to provide a general definition of 'stem cell:' a set of individually necessary and jointly sufficient conditions that pick out all and only the cases to which the term applies. Zipori proposes plasticity as such a condition. Having many possible developmental outcomes, he claims, is a property of all stem cells that is "not shared by any other cell type" (2009, 166). But there is already a widely-accepted general definition: stem cells are capable of self-renewal and differentiation. To simply stipulate an alternative is otiose. So the plasticity definition needs some additional motivation.

Zipori finds this motivation in the wide gap between the abstract general definition of 'stem cell' and experimental data produced by stem cell experiments (Chapter 2). The former connects to the latter only via additional representational assumptions and specification of key parameters, including organismal source, characters of interest, temporal duration, and methods of manipulating cell environments. As noted above, the stemness alternative is explicitly opposed to the collinear model, interpreted as a strict, general theory of cell development (§4.2). In that general theory the relations linking cell compartments are laws of development, specifying the ineluctable order of developmental processes, with cell and organismal levels closely coupled. The branching pattern of cell development in multicellular organisms is linear, irreversible, and follows a fixed sequence of stages. Any cell that can self-renew and give rise to more differentiated cells within this rigid developmental hierarchy counts as a stem cell.[77]

Proponents of stemness oppose this theory, and the general definition of 'stem cell' associated with it, on grounds that it makes exact predictions about patterns of cell development that do not fit our experimental data. This is a powerful motivation; rather than a superfluous new definition of 'stem cell,' the stemness alternative is intended to address empirical inadequacy of an existing theory. It is not clear, however, that anyone actually endorses this general hierarchical theory of cell development instead of a more relaxed, exception-permitting counterpart. If relations in the model are *not* taken to represent strict empirical regularities that describe changes in cell character values over time, no disconfirmed predictions follow; the charge of empirical inadequacy does not stick.

Few biologists would claim that cell development is strictly hierarchical and irreversible. For one thing, strict empirical regularities are extremely rare in biology. Moreover, the prevailing definition of 'stem cell' is perfectly compatible with exceptions to linear pathways of cell development. The general theory sketched in the previous paragraphs is just one elaboration of the abstract stem cell model; only one of many ways to fill the gap between the prevailing general definition and evidence from concrete experiments. Relations in the abstract stem cell model need not be interpreted as representing strict empirical regularities about the temporal order of cell traits. And, given the paucity of such regularities throughout biology, it would be odd for stem cell biologists to opt for this interpretation. These considerations suggest that the stemness alternative is aimed at a straw target.

Aspirations to generality are also revealed in Zipori's objections to a *universal* stem cell signature. These objections indicate conceptual and empirical problems for a theory based on the prevailing stem cell concept, which identifies a set of character values distinguishing stem cells from all other cell types, irrespective of source or method. Zipori's plasticity definition is designed to avoid these problems (§4.3). His treatment of self-renewal is particularly revealing in this regard: because there is no general way to characterize this capacity independently of experimental methods and contexts, self-renewal must be omitted from the general definition of stemness. But another option is to interpret stem cell signatures as *local*; that is, relative to an organismal source, characters and duration of interest, and methods used to manipulate cells. Such local signatures are revealed, albeit provisionally, by stem cell experiments. Zipori's critique presupposes that stem cell biology aims not only to reveal and describe a collection of local stem cell signatures, but also to confirm a theory of general cell development independent of experimental variables. But, for the most part, stem cell biologists evince scant interest in such theories. This explains their tepid response to the stemness alternative. In an experiment-driven field, this general definition is a solution in search of a problem.

4.6   Evidence for plasticity

By the same token, in an experimental field, empirical adequacy takes precedence in adjudicating between competing theories or models. A second role for the stemness alternative is to fit the experimental evidence. Its proponents argue that two kinds of experimental data support their view over the prevailing stem cell concept: evidence of adult stem cell plasticity, and results of recent attempts to identify a genetic signature for stem cells. However, these evidential arguments do not succeed. I criticize each in turn.

The thesis of adult stem cell plasticity has two parts: first, tissue-specific stem cells have greater differentiation potential than required by their tissue localization and, second, cell development is not irreversible, in that cells need not proceed in one direction along a linear developmental pathway, from higher to lower levels of differentiation hierarchy. Although both claims follow from the stemness alternative (see §4.1), the degree to which they are supported by experimental evidence is a matter of long-standing controversy. The controversy has several roots. First, as discussed in Chapter 3, single stem cells are not measured directly in these experiments, making interpretation of results difficult. In addition, the debate is complicated by political interests. If stem cells are all pluripotent and interchangeable (as the stemness alternative entails), then stem cells derived from adult organisms could substitute for embryonic stem cells, without loss of potential. Opponents of embryonic stem cell research have repeatedly argued that hoped-for therapeutic benefits can be achieved without destroying human embryos. Interchangeably 'plastic' stem cells are a highly welcome result for some, less welcome for others. Though no scientists actively seek destruction of human embryos, many have invested time, effort, and funds in embryonic stem cell research and are reluctant to abandon the ongoing projects on which they have staked their careers. Many would also resist conforming their research to moral prescriptions they do not endorse. So many different interests come into play. Scientific debate about plasticity is repeatedly exploited by political opponents of embryonic stem cell research, making for muddy evidential waters.

To date, evidence for pluripotent adult stem cells is uncertain, disputed, and tied to specific experimental procedures. All the uncertainties discussed in previous chapters therefore apply to experiments purporting to identify pluripotent stem cells in adult organisms. Many of these experiments face additional criticism for failing to meet the single-cell standard (see Chapter 3). Claims of adult stem cell plasticity are therefore equivocal at best. The contested nature of the evidence undercuts claims to greater empirical adequacy by proponents of stemness. More decisively, the uncertainty principle exposes internal tensions in stemness account. The first part of the plasticity thesis asserts that the differentiation potential of a single cell in a stem state exceeds that cell's realized developmental output. This is predicted by the stemness view; by definition, a cell in the stem state can give rise to a wide array of differentiated cell types. Which outcome is realized depends on the cell's environment, or niche. But, at the single-cell level, the only way to test for stem cell capacities is to place a cell in a specific environment and measure its developmental output. Given a homogeneous population of stem cells, it is possible to test for reproductive capacities in a range of environments. So confirmation of stem cell plasticity depends on a homogeneous stem cell population. *Yet the stemness account denies that such populations can be identified and isolated.* Experimental support for the stemness account hinges on a homogeneity assumption that is necessarily unsupported, according to that very account.[78] This internal tension obviates evidential support for the stemness alternative from adult stem cell plasticity experiments.

Finally, even if one grants the 'plasticity interpretation' of the experimental evidence in these cases, this does not decisively favor the stemness account over the prevailing view. These data, so interpreted, are incompatible with models of cell development that define levels of cell hierarchy in terms of irreversible commitment to a stable differentiation pathway. But the irreversible, exceptionless developmental hierarchy is a straw target. To assume that we must choose between two 'absolute' models of cell development – orderly progression of hierarchical levels, or a field of open possibilities – is to endorse a false dichotomy.[79] Arguments that self-renewal and production of a stable hierarchy are not absolute requirements for stem cells, because exceptions to each have been observed, fall similarly flat (e.g. Zipori 2004; Mikkers and Frisén 2005). There are no absolute requirements for stem cells in this sense, only evidence for stem cell capacities in diverse experimental contexts.

The second kind of evidence is very different, obtained with new methods of global gene expression analysis and transcriptional profiling. These high-throughput methods are associated with the emerging interdisciplinary field of systems biology (see Chapter 9). At the turn of the twenty-first century, these cutting-edge techniques were used to seek a shared molecular signature in cell populations from different stem cell experiments and organismal sources. Two large-scale comparative studies were published in 2002, each reporting 200–300 genes more highly expressed in stem cells than other cell types. The large number of differentially-expressed genes indicated that the signature of stemness was complex rather than governed by a few genes. This result was not surprising, however, or seen as particularly problematic, though a simple signature would have been welcomed. But problems emerged when results of these and a third study were compared (Fortunel et al. 2003). Only *one gene* was identified by all three as a signature of stemness. The lack of overlap cried out for explanation and the debate played out quickly on the editorial pages of *Science*, with the authors highlighting differences in their experimental methods which might account for the discrepancies.

Zipori and colleagues propose another explanation: the experiments did not reveal a shared pattern of gene expression for all and only stem cells because there is no such pattern. If gene expression is the determinant of a cell state then there is no unique molecular signature for stemness, and no common genetic program for all and only stem cells. The stem state is, instead, multiply realizable by diverse gene expression patterns. This argument takes results of stem cell experiments to directly reflect underlying molecular reality. But it does not follow, from the fact that results of experiments aimed at discovering the molecular signature of stem cells are diverse, that there is no unique molecular signature for stem cells. Different stem cell experiments use different parameters and assumptions to flesh out the abstract stem cell concept. Diverse results across experimental contexts are therefore to be expected. Within a particular experimental context, with fixed assumptions and parameters, heterogeneity in results could reflect diversity among cells used in the experiments. Eliminating this possibility requires isolating pure populations of stem cells to use as input for gene expression studies. The conclusions about stem cells inferred from the resulting data are only as well-supported as the premise that stem cell populations serving as input are pure populations. For the 2002–2003 experiments, singly and collectively, this premise is dubious. Overall, then, the available evidence does not decisively support stemness over the prevailing view. The stemness alternative does not fulfill this second role, as a source of empirically accurate predictions.

4.7 Molecular genetic explanation

The above considerations show that the stemness account cannot provide a general definition that fits all the experimental evidence. The trade-off between these two aims puts the stemness account at odds with itself. Cartwright (1983) describes this trade-off as between explanation and empirical accuracy. In physics, from which Cartwright's cases are drawn, explanation often involves subsuming diverse phenomena under a general theory. But in experimental biology, most explanations refer to underlying molecular mechanisms. Such mechanisms include molecular characters of cells, including surface molecules, intracellular proteins, RNA, chromosomes, and DNA. A common pattern of explanation in experimental biology is to account for cell morphology and function in terms of these and other interacting molecules. A third role of the stemness alternative is to set the stage for molecular explanations of the stem state. In this role, the account succeeds.

The idea of molecular explanation does not immediately follow from the stemness view, which merely suggests that we can harness stem cell capacities by learning the rules that govern transitions from one cell state to another. But proponents of stemness hold that these rules are to be found at the molecular level. Blau and colleagues (2001) speculate that the functional stem state is realized by cells with diverse patterns of gene expression, i.e. different cell types. Quesenberry's chiaroscuro theory includes the hypothesis that chromatin structure changes throughout the cell cycle, causing patterns of gene expression to change, thereby revealing or masking stem cell capacities. Mikkers and Frisén (2005) suggest that the molecular signature of stemness consists of mechanisms needed to resist the force of development. Zipori's proposal, the most detailed, is that gene expression patterns explain cell development. All these accounts emphasize the molecular ground of stem cell capacities. So the concept of 'cell state' has two aspects: function at the cellular level and pattern of gene expression at the molecular level. The latter is thought to explain the former. Zipori (2009) explicitly connects the two, claiming that "the stem state is characterized by pluripotency that entails genome-wide gene expression" (2009, 201).

The assumption that cell states have a molecular ground is minimal, compatible with many possible relations between stemness and molecular traits of a cell. The simplest relation is control by a single molecular element: a 'master gene' expressed in all and only cells in the stem state. But such simple relations between cell and molecular traits are exceptional. The molecular basis of most cell traits is more complex. Another possibility is that one set of interacting molecules (genes, proteins, RNA) underlies all stem states. A general explanation of stemness would then describe these molecules, their interactions, and arrangement: the molecular mechanism for stemness. In principle, the stemness mechanism could be entirely internal to the cell ("cell-autonomous") or extend beyond the cell membrane to include features of the cell's environment.[80] Experimental evidence supports the latter. There is very strong evidence that a cell's microenvironment – its niche – determines its course of development. However, niche effects on cell *capacities*, as opposed to their realization, are difficult to confirm owing to the uncertainty problems discussed in §4.6 and Chapter 3.

Further, Zipori and colleagues argue that current evidence tells against the idea that all stem cells share a common underlying mechanism for stemness. Instead, stem cells from different sources or in different environments use quite different molecular mechanisms to realize the stem state. This explains the minimal overlap across gene expression studies (§4.7). There may not be any set of traits shared by all and only stem cells from any organismal source. Hence "unveiling the molecular programs for such features is unlikely to provide the full picture of what makes a stem cell a stem cell" (2005, 2718). But, here, Zipori and colleagues underestimate the power of molecular explanations. The current evidential situation suggests that there is no *simple and general* molecular explanation for stem cell capacities. But explanations that engage the complexity and variability of stem cell traits offer a promising way forward. The next chapter examines explanations of this sort, termed 'mechanistic explanations.'

4.8   Conclusions

The stemness account is proposed as an alternative to the prevailing stem cell concept, explicated in the abstract model. But it suffers from an excess of ambition, aiming to be a general definition that fits all the experimental evidence and sets the stage for molecular genetic explanation. These three desiderata – generality, empirical adequacy, and explanation – are all scientifically important, but in tension with one another, such that no single theory (or model) can satisfy all three at once. So it is no surprise that the stemness account does not succeed. The prevailing stem cell concept manages these trade-offs more effectively. The abstract stem cell model is general and unifying, and conforms to evidence of stem cell capacities relative to particular experimental methods. The stemness critique is directed not against this model, but against a general stem cell theory that posits a universal stem cell signature.

This result explains the lack of response to the stemness critique by most stem cell biologists. Stemness critiques miss their target in a peculiar way. The argument is that there is a general theory of development, which recent evidence shows to be incorrect. We therefore need another theory, consistent with the evidence. This is classic theory replacement in light of new evidence. But the critique goes astray in assuming that there is a general theory of development to start with. Stem cell biology does not have one and, more importantly, does not need one. The target of the stemness critique is a shadow cast by mid-twentieth century philosophy of science.

In the matter of explanation, however, the stemness account makes a genuine contribution, setting the stage for molecular explanation of stem cell capacities. The "state or entity?" debate is better understood as being about how to design experiments that provide evidential support molecular explanations, rather than which theory is correct. And, here, questions of generality resurface. The prevailing research program makes progress by collecting many local signatures, through diverse experiments that only share a minimal stem cell concept and basic methodological structure. How do these fragmented experimental results support robust models or explanations? The chapters in Part II examine this question, with philosophical accounts of biological explanation coming to the fore.

Part II

5
Mechanistic Explanation: The Joint Account

5.1 Introduction

After stem cell phenomena are characterized, scientists seek to explain them in terms of underlying molecular mechanisms. This style of explanation, which compiles results of many experiments, is widespread in biology and medicine. Mechanistic explanations describe the parts and processes underlying phenomena, such as DNA synthesis, cell metabolism, memory, and stem cell capacities, differentiation, and self-renewal. Philosophical accounts of mechanisms and mechanistic explanation largely concur on their essential features. There is less consensus, however, on what makes a mechanistic description *explanatory*. Scientific explanation is traditionally associated with 'covering laws' that subsume many diverse phenomena. But laws of this sort appear rarely, if ever, in mechanistic explanations of biological phenomena. More promising is the idea that mechanistic descriptions explain in virtue of correspondence to real causal relations among working components of a mechanism. However, despite its plausibility and rigorous explication by a new theory, this causal account is not fully satisfactory either. This chapter builds on earlier philosophical work to propose a new account of mechanistic explanation, centered on the concept of jointness.

The term 'jointness' refers to interdependence among parts of a mechanism. It is a well-known feature of mechanisms that their components work together to bring about phenomena that do not result from the same components in isolation.[81] The proposal here is that jointness, the working together of parts, is at the core of mechanistic explanation. This 'joint account' of mechanistic explanation has several advantages over the main alternatives, which attribute explanatory power to laws and causes, respectively. It resolves tension between reductive and non-reductive aspects of mechanistic explanation, explicates the sense in which these explanations are multi-leveled, and distinguishes them from causal explanations *simpliciter*. The joint account also has important consequences for the role of genes in mechanistic explanations of organismal and cell development, examined in the next chapter.

Three caveats should be noted at the outset. First, the rich and complex philosophical history of the term 'mechanism' is not addressed. The joint account concerns practices of twentieth- to twenty-first-century life science, which bear few explicit traces of seventeenth-century mechanical philosophy or eighteenth-century opposition to vitalism. Continuities with earlier ideas are not explored here. Second, the joint account is not proposed as a general theory of scientific explanation. The aim of this chapter is to clarify one important kind of explanation, not to rule out others. Finally, the joint account is defended here only for "experimental biology:"[82] fields that investigate biological systems by manipulating their components in laboratory experiments. These include molecular biology, cell biology, neuroscience, biochemistry, developmental biology, some areas of genetics, and (plausibly) biomedical fields such as immunology and stem cell biology. All these fields aim at mechanistic explanations. Whether or not other fields (such as evolutionary biology) also aim at these explanations does not bear on the argument here. The joint account is not intended to be general or to exhaust the meaning of 'mechanism,' but rather to capture important features of explanations in experimental biology, including stem cell research.

5.2 Mechanisms

Philosophers have proposed a number of definitions of 'mechanism.' For example:

> Mechanisms are entities and activities organized such that they are productive of regular changes from start or set-up to finish or termination conditions.
>
> (Machamer et al. 2000, 3)

> A mechanism underlying a behavior is a complex system which produces that behavior by the interaction of a number of parts according to direct causal laws.
>
> (Glennan 1996, 52)[83]

> A mechanism is a structure performing a function in virtue of its component parts, component operations, and their organization. The

orchestrated functioning of the mechanism is responsible for one or more phenomena.

(Bechtel and Abrahamsen 2005, 423)

These definitions concur on the essential features of a mechanism. First, all emphasize the causal, productive character of relations among components. The "entities and activities" of Machamer et al. correspond to Bechtel and Abrahamsen's "parts and operations," and Glennan's "parts and direct causal laws." Organization is another essential feature. Causally-active components are organized so as to rise to an entire sequence of "regular changes from start to...finish" (Machamer et al.), an overall "behavior" (Glennan), or 'one or more phenomena' (Bechtel and Abrahamsen). Though the concepts of cause and organization both need further analysis, there is strong consensus that a mechanism is a complex causal system with multiple components, which interact to produce some overall phenomenon.

Phenomena of molecular biology, cell biology, and neuroscience are often explained in terms of underlying mechanisms (Bechtel 2006, Craver 2007, Darden 2006). The same is true for stem cell biology. For example, consider the mechanism of stem cell self-renewal and differentiation in testes of the fruitfly *Drosophila*, currently one of the best-characterized in the field. The testis is a relatively simple organ, which contains germline stem cells (GSC) that give rise to sperm. Cells, genes, and molecules involved in *Drosophila* development have been described in great detail. GSC are unipotent stem cells which engage in the simplest developmental process involving both stem cell capacities: asymmetric cell division. Altogether, the GSC mechanism in *Drosophila* testes is one of the simplest and most complete in stem cell biology today. A brief, simplified description illustrates the key features of mechanistic explanation (Figure 5.1).[84]

During fly development, the testis forms as a tube with a clump of somatic cells at one end: the 'hub.' Approximately ten cells (GSC) attach to the hub via adhesion molecules expressed on cell surfaces. These surface molecules are attached, in turn, to proteins that span the cell membrane, allowing chemical signals to pass from hub cells to GSC. Signaling proceeds via a sequence of minor chemical changes to proteins that alter their structure and function (details omitted, for brevity). Within a cell, signals initiate cascades of further chemical reactions. The GSC signaling pathway, termed 'Jak-STAT,'[85] emanates only from the hub. So GSC are only found at this end of the testis and their number limited to about ten at any given time. Other testis cells (including hub cells themselves) make a protein that inhibits the Jak-STAT pathway.

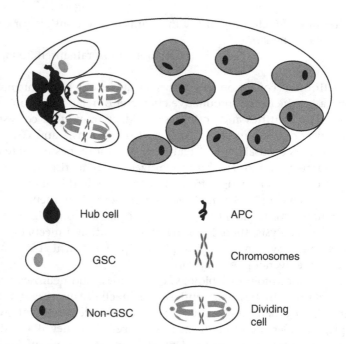

Figure 5.1   Mechanism of asymmetric GSC division in *Drosophila* testis

When a hub-attached GSC divides, its chromosomes condense, replicate, and line up in the center of the cell. At each end of the cell, a centrosome or 'spindle' acts as a pole, drawing one copy of each chromosome toward itself. The two centrosomes define the plane of division along which the GSC splits in two. A set of proteins termed 'APC' anchor the centrosome near the hub-GSC boundary. When cell division begins, the centrosome duplicates and the new copy migrates to the opposite end of the cell. The plane of division is therefore such that one progeny cell remains attached to the hub, while the other is physically separated from it. Separation from the hub interrupts Jak-STAT signaling, initiating differentiation. Differentiating germ cells migrate toward the other end of the testis, eventually forming sperm. So each division produces one new GSC attached to the hub like the parent, and one differentiating cell, which begins a new sequence of interactions and changes, culminating in spermatogenesis. Such is the (simplified) mechanistic description of stem cell capacities in *Drosophila* testes.

The GSC example illustrates the key features of mechanisms, according to the consensus view. The mechanism consists of many diverse parts: molecules, cells, proteins, and subcellular structures (see Figure 5.1, bottom). These material parts engage in a variety of causal activities, including binding, phosphorylation, migration, division, and replication. Together, all these tightly-orchestrated interactions yield an overall phenomenon: GSC self-renewal and differentiation. The distinction between parts and overall phenomenon reveals the hierarchical, or multilevel, structure of mechanisms. Mechanisms of interest for experimental biology include (at least) two such levels.[86] But the notion of a mechanistic level remains somewhat mysterious; it cannot be defined in terms of size, containment, mereology, or biological organization.[87] The first and last points are clearly shown by the GSC case. These components run the gamut from cells, small molecules, macromolecules (proteins and DNA), and subcellular structures. Chemical bonds, cell–cell interactions, arrangement of chromosomes, and even the geography of the testis, are included. The example also highlights the importance of spatio-temporal localization for components.[88] The entire mechanism is contained within the testis, an organ. Within this organic context, components' causal relations dovetail with their spatio-temporal position. For example, stages of cell division interlock with relative positions of cell types and the operation of molecular signaling pathways. The description traces all these orchestrated interactions, verbally and diagrammatically.[89] Key details include which components interact with one another, in what order, and in what position with respect to the entire mechanism.

With these considerations in mind, the consensus view can be stated as follows:

(M) A mechanism S consists of multiple diverse components (x's) engaging in causal relations or activities (ϕ's) such that x's and ϕ's are spatially and temporally organized so as to produce some overall phenomenon.[90]

Like the biological mechanisms that furnish its examples, (M) lends itself to diagrammatic representation (Figure 5.2a). Furthermore, a working component of a mechanism (x ϕ-ing) may itself be a complex system of causally interacting parts – a mechanism in its own right (Figure 5.2b). Similarly, a mechanism S may be a component of a still higher-level mechanism. The biological world, as we understand it, is rife with hierarchies of this sort. The next task is to understand how description of a mechanism amounts to an *explanation*.

Any explanation consists of an explanandum that is explained, an explanans that does the explaining, and a relation connecting the two.

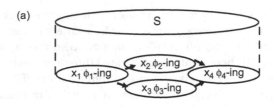

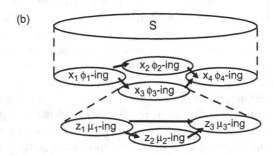

Figure 5.2   The consensus view of mechanisms (after Craver 2007, and Craver and Bechtel 2007). (a) Two levels: component x's and overall mechanism S. (b) Downward expansion: three mechanistic levels

The consensus view (M) suggests that the explanatory structure of a mechanistic explanation (hereafter MEx) mirrors the hierarchical structure of the mechanism it describes. That is, the explanandum is a description of the overall phenomenon of interest (S) and the explanans a description (verbal, diagrammatic, or both) of organized interacting parts (ϕ-ing x's). The next sections show that MEx, so understood, does not conform to either of the two main philosophical accounts of scientific explanation: law-based or causal.

5.3   Law-based explanation

The traditional view is that scientific explanation consists in deriving a statement to the effect that a phenomenon of interest occurs (the explanandum) from a general law and initial conditions (the explanans). On this classic account, logical derivation is the connecting relation, and the epistemic role of explanation is to show that a phenomenon of interest is to be expected. Laws are traditionally defined as universal exceptionless generalizations of wide scope. Explanations invoking such laws subsume a phenomenon of interest to the wider domain 'covered' by the law. Marcel Weber (2005) applies this covering-law view to MEx, arguing that these explanations rest on physical and chemical principles about the behavior of molecules. In the domain of experimental biology, he claims, only physicochemical principles, which hold under an extremely wide range of conditions, qualify as "*genuine* laws of nature" (2005, 34). The rest of a mechanistic description merely "states how the physicochemical theory should be applied," providing initial and boundary conditions such that descriptions of higher-level phenomena can be derived from lower-level laws (2005, 25–26). Weber's account of MEx thus retains the core ideas of the traditional covering-law theory of explanation.[91] On his view, MEx rest on invariant, stable generalizations about molecules, from which many predictions about higher-level phenomena (cells and organisms) can be derived. Physicochemical laws explain by unifying, in this sense, the many and varied phenomena of experimental biology.

Weber's law-based account has the advantage of clearly explicating the 'bottom-up' direction of MEx. It also makes sense of the tendency of mechanistic descriptions in biology to "bottom-out" at the molecular level, suggesting a fundamental explanatory role (Machamer et al. 2000). But these strengths are offset by a major weakness: laws are "peripheral" to MEx in experimental biology.[92] Universal exceptionless generalizations are conspicuously absent from mechanistic descriptions like the GSC case (§5.2). This is not to say that MEx contradict any physicochemical laws, just that such laws are seldom included in descriptions of biological mechanisms. Their absence manifests in multiple ways. For one, MEx allow for exceptions. To return to the GSC example, only about 80% of GSC divisions have a plane of division parallel to the hub-GSC binding site. Though the GSC mechanism only works as described in ≤80% of cases, the MEx is still acceptable. Second, general and specific descriptions do not play different roles in the MEx, analogous to laws and initial conditions. Some parts of the GSC MEx describe very general features, such as arrangement of chromosomes and subcellular structures in dividing cells (see Figure 5.1). Others, such as the interaction of Jak-STAT signaling, hub cells and APC position, may be restricted to one organ of a single species. The simplest and most perspicuous way to explain biological phenomena with fundamental laws would be to articulate the relevant laws, list the initial and boundary conditions, and exhibit how explanandum-phenomena result. But the mechanistic description of GSC, for example, does not begin with a general account of cell division and proceed to add more specific conditions until self-renewal and differentiation are accounted for. Instead, general and specific features of the case are mingled. This example is typical; the structure of MEx in experimental biology does not reveal an explanatory dichotomy of physicochemical principles and all other features of a mechanistic description.

Third, in many MEx the lowest level does not describe all and only molecular components, but spans multiple levels of biological organization. The GSC example above is typical in this regard as well: the component-level description features not only molecular structures and relations, but also cells, organelles, macromolecular complexes, and organs.[93] This undercuts Weber's insight about the molecular "lower bound" of MEx. Finally, the explanandum–explanans relation for MEx is not plausibly represented as logical derivation. At least in experimental biology, these explanations lack the elegant simplicity of derivations, bristling with detail about parts of organisms and their intricate causal and spatio-temporal organization. Nor do they fulfill the roles of derivational explanations; MEx in experimental biology do not secure predictions or subsume many diverse phenomena. Overall, they exhibit none of the key features of law-based explanation. Law-based accounts such as Weber's therefore appear as Procrustean attempts to assimilate MEx to more traditional theories of explanation.

A defender of law-based explanation might respond by noting that MEx for most (if not all) biological phenomena are incomplete. Their wealth of non-molecular detail and paucity of laws might merely be symptoms of immaturity. Perhaps law-based MEx are a regulative ideal, guiding our partial explanations of biological phenomena toward completion. Then, as young fields of experimental biology progress (if they do), we should see simplification, streamlining of non-molecular details, and greater emphasis on physicochemical principles. It is difficult to discern general trends across the many biological fields that use MEx. But history of biology offers little support for the idea of laws as a regulative ideal for MEx. Twentieth-century projects focused on single genes, master molecules, and simple models of regulation are gradually being superseded by cellular, organismal, and system-level approaches. There has not been a marked increase in explanatory use of laws from physics and chemistry, despite attempts since the nineteenth century to make the latter foundational for biology (Keller 2002). Instead, continued research on mechanisms underlying biological phenomena tends to increase what Weber considers non-explanatory details.

Perhaps the clearest indication is the trend toward expansion rather than contraction of levels in MEx. On the law-based account, everything above the molecular level merely specifies how general physicochemical principles apply in a particular case of interest. Biologists should therefore seek ways to streamline multilevel mechanistic hierarchies, not build more elaborate ones. But, in fact, the opposite occurs. The trend in neuroscience, genomics, and stem cell biology is toward more detailed "multilevel explanatory frameworks" (Craver 2007). The peripherality problem is a decisive objection to law-based accounts of MEx. This motivates the main alternative to law-based explanation that to explain a phenomenon is to reveal how it fits into "the causal structure of the world" (Salmon 1989).

5.4   Causal explanation

5.4.1   The manipulability theory

There are many theories of causality. However, in philosophy of biology, and for MEx in particular, James Woodward's (2003) manipulability theory predominates. This theory analyzes causality as a relation between values of variables, X and Y. X causes Y if and only if there is a possible manipulation of some value of X, under idealized experimental conditions, such that the value of Y changes. Woodward's is a counterfactual account of causality, hinging on what *would* happen to the values of variables in an idealized experiment, or *intervention*. An intervention I on variable X with respect to variable Y is a causal process that determines the value of X in such a way that, if the value of Y changes, then the change in Y occurs only in virtue of the change in X.[94]

The concept of an intervention provides a regulative ideal for experiments aimed at discovering causal relations: approximate the conditions of ideal experimental interventions. Insofar as they meet this standard, experiments reveal genuine patterns of counterfactual dependence among sets of entities and their properties, i.e. causal relations in the world. Because invariance is required only under *some*, not all, interventions, causal explanations need not include general laws. The range of interventions under which a dependency relation between values X and Y holds may be broad or narrow. Within their range, 'fragile' dependency relations are no less causal than those with a much wider range of invariance: universal causal laws.[95] Experiments, if correctly designed, reveal relations between values of variables that are invariant under some interventions. The relevant counterfactuals for causal relations are tested by realizing particular values of X and observing the values of Y under controlled circumstances that fall within the range of invariance for X and Y. Woodward's theory thus offers an appealing analysis of causal relations revealed by experiment, including those described in MEx (Craver 2007, Glennan 2002, Woodward 2002). Hereafter, the term 'cause' refers to the concept analyzed by the manipulability theory.

5.4.2   The causal-mechanical account

Carl Craver extends Woodward's manipulability theory to account for norms of MEx and explicate the relation between mechanistic levels. Using detailed case studies from neuroscience he defends a "central criterion of adequacy" for MEx: "account fully" for the explanandum-phenomenon in terms of (i) component entities whose real existence is justified by multiple criteria; (ii) causal relations (activities) validated by experiment; and (iii) causal, spatial, and temporal organization of entities and activities (2007, 139).[96] An account is "full" if and only if it describes all *relevant* components of a mechanism. Causal relevance is in turn analyzed in terms of "mutual manipulability," (2007, 152–160).[97] A working component (x ϕ-ing) is relevant to an overall mechanism's behavior (S Ψ-ing), if the latter can be manipulated by intervening on x's ϕ-ing *and* x ϕ-ing can be manipulated by intervening on S Ψ-ing. A component is irrelevant to a mechanism if neither x's ϕ-ing nor S Ψ-ing can be manipulated by intervening on the other. These two sufficient conditions link the hierarchical structure of MEx to experimental practices in neuroscience, which employ both 'top-down' and 'bottom-up' strategies to detect components of mechanisms.

Craver's "causal-mechanical account" of MEx dispenses with the requirement for covering-laws. Instead, mechanistic descriptions have explanatory power in virtue of correspondence to causal relations among actual components of a real mechanism. But the multilevel aspect of MEx is also crucial. Craver does *not* fully define the relation between mechanistic levels in terms of manipulation and intervention; conditions for mutual manipulability are sufficient only. On his account, MEx are "constitutive," rather than causal, explanations. But it is easier to say what constitution is not, than what it is. In biological mechanisms, this relation is "non-aggregative," "non-decomposable," and distinct from temporal order, size, physical containment, or mereology (Bechtel and Richardson 1993, Craver 2007, Wimsatt 2007).[98] Though Craver's account emphasizes the hierarchical structure of mechanisms and MEx, his causal explication leaves some questions unanswered.

5.5   Three problems

Craver's is the best-developed account of MEx in biology currently on offer. But it suffers from three problems. First, the explanandum is ambiguous.

Second, the causal-mechanical account omits the 'direction' of MEx, from parts to whole. Third, it leaves questions of modularity in biological mechanisms unresolved. All three problems concern the relation of causality to hierarchical structure of mechanistic levels. Together, they motivate a new account of MEx.

5.5.1 Ambiguity

It is widely-recognized that there are no mechanisms *simpliciter*, but only mechanisms *for* some phenomenon X. The phrase 'mechanism for X' marks the target of explanation, which the explanandum describes. In experimental biology, 'X' usually refers to a complex behavior or function, such as glycolysis, memory, or cell division. However, 'mechanism for X' can refer either to a mechanism's working or to effects thereby produced – the mechanism's outcome. Such process/product ambiguity is prevalent in biology, where the same term commonly refers to both; important examples include growth, reproduction, infection, and death.[99] So the explanandum for MEx is ambiguous: is it the mechanism's working (S Ψ-ing) or phenomena caused by its working [downstream effect(s) P]? The manipulability theory suggests the latter; Craver's own account the former.

To see this, it is helpful to consider separately the causal relations at different mechanistic levels (Figure 5.3). At the higher level an overall mechanism S works (Ψs) and thereby produces phenomenon P; more succinctly, S Ψ-ing causes P (Figure 5.3a). Causal claims of this sort are exceedingly common in biology; cell division produces two cells, long-term potentiation increases synapse strength, and so on. On Woodward's theory, such statements mean that an intervention on M Ψ-ing changes the value of P. However, this does not suffice for MEx, which also describe causal relations at the component level. At the component level, spatio-temporally localized parts of the mechanism (x's) play causal roles (ϕ-ing) such that a change in any x's ϕ-ing makes a difference to some other component(s) (Figure 5.3b). In the GSC example, x's are specific molecules, cells, proteins, and subcellular structures, and ϕ's are causal activities, such as binding, phosphorylation, migration, division, and replication. Connecting the two levels as cause-and-effect takes the form of a dependency relation between x ϕ-ing and P that specifies the difference that intervening on the former makes to the latter (Figure 5.3c). The two levels – upper and lower – offer alternative causal explanations of P. However, Craver's characterization of MEx as "constitutive" points to the working mechanism itself as the target of explanation (Figure 5.3d). This tension between the manipulability theory and the multilevel aspect of MEx renders the explanandum in Craver's account ambiguous.

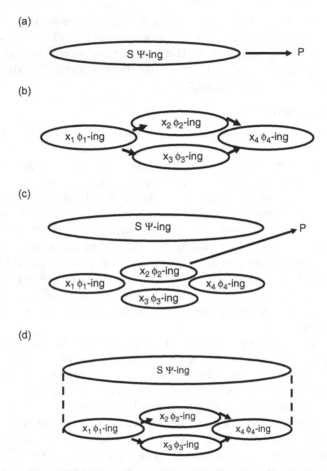

Figure 5.3 Levels of causal relations in MEx. (a) System-level: mechanism S Ψ-ing produces phenomenon P. (b) Component-level: component x's causally interact (φ-ing) to effect one another. (c) Inter-level: component φ-ing x's make a difference to S's outcome (P). (d) Constitutive MEx: component x's interact (φ-ing) to constitute mechanism S's Ψ-ing

5.5.2 Direction

A second problem is that explication of interlevel constitution in terms of causal relations (mutual manipulability) does not account for the bottom-up direction of MEx, a basic feature of this mode of explanation (see Chapter 4). Mutual manipulability requires both top-down and bottom-up dependency relations to establish or rule out causal relevance of a component (see §5.4.2). This reflects the design of experiments aimed at discovering mechanisms and testing MEx. An overall system may be manipulated to reveal effects on its parts, or a part may be manipulated to reveal effects on its overall mechanism. Woodward's theory offers a satisfying analysis of these practices. However, extension of the manipulability analysis to the interlevel relation, representation of which connects explananda and explanans in MEx, does not account for the part-to-whole direction of these explanations. *Mutual* manipulability, as the name implies, cuts both ways. Though these conditions explicate the regulative ideal of causal relevance, they leave obscure why a description of components explains a higher level mechanism (or its effects), rather than the direction of explanation running both ways. If Craver's account is correct, scientists would do just as well to look for "overarching" as underlying, mechanisms.

One response is that, in practice, experimental biologists *do* look both 'up and down' mechanistic levels. Indeed they do – when *discovering* mechanisms. Much philosophical work on biological mechanisms is oriented toward discovery, clarifying the strategies and heuristics scientists use to detect and construct models of real-world mechanisms.[100] This approach yields accounts such as Craver's, that articulate norms for experimental practices that integrate different mechanistic levels in a single explanatory framework. But such accounts do not tell us how MEx, once constructed, explain higher levels in terms of lower ones. Craver's view of the matter is complicated by his thesis that causal relations do not cross mechanistic levels (Craver and Bechtel 2007). Arguments for this thesis are intuitive, but persuasive. If the interlevel relation in mechanisms were simply causal, then intervening on a mechanism M would cause a change in one or more components along a pathway $I{\to}M{\to}x$, and intervening on one or more components would cause a change in M along a pathway $I{\to}x{\to}M$. But, intuitively, an intervention on an overall mechanism does not *cause* a change in one or more components. To return to the GSC example: disrupting Jak-STAT signaling inhibits asymmetric GSC division, not because the former *causes* the latter, but because a change in the former *is* a change to the overall mechanism. The interlevel relation that is the crux of mechanistic descriptions does not reduce to cause-and-effect. Rather, such an intervention *is* a change in one or more of its components and *vice versa* ($I{\to}M$ iff $I{\to}x$). Accordingly, Craver's mutual manipulability conditions do not define or fully analyze the constitution relation.[101] But all that is said of this relation is in terms of causal dependence, which does not capture the bottom-up direction of MEx. As with the explanandum, Craver's account exhibits tension between the hierarchical and causal aspects of MEx.

5.5.3   Modularity

A third problem concerns constraints on a mechanism's components as represented in MEx, entailed by the manipulability theory. These constraints take the form of *modularity theses*. 'Modularity' here refers to independence among a mechanism's components: a component of a mechanism is modular if and only if it is possible, in principle, to intervene on its activities independently of other components of the mechanism. Modular components can be independently experimentally manipulated.[102] Woodward (2002) proposes modularity as a constraint on components of a mechanism as represented in MEx. However, this constraint may be interpreted in several ways.[103] The weakest modularity thesis states that for x to be a component of a mechanism S, it must be possible in principle to intervene on x's activities independently of other components. More precisely:

Mod-1: If x is a component of mechanism S, then there is a conceivable intervention on x that changes its φ-ing independently of other components of S.[104]

Mod-1 is a very plausible constraint on MEx. If we cannot conceive of changing any of a component's properties or behaviors independently of other constituents of a mechanism, then why conceive it as a distinct component at all? If we can conceptually individuate a working component of a mechanism (x φ-ing), then we can imagine an experiment that 'surgically' intervenes on x to change its φ-ing without altering other components. A simple example would be to remove x from the rest of the mechanism and observe the effects. Description of such thought-experiments for each part of some mechanism S yields a 'parts list' for S. But such a list is not sufficient for MEx.

If a component of mechanism S is modular in the sense of *Mod-1*, then the generalization describing its productive activity (x φ-ing) can, in principle, change independently of generalizations describing productive activities of other components (Woodward 2002). Generalizations of this sort describe the results of hypothetical experiments that isolate or 'surgically manipulate' components of S. A second, stronger modularity thesis relates the conceptually separable activity of a component to the overall mechanism via a modular causal generalization:

Mod-2: If x is a component of mechanism S, then there is a modular causal generalization relating x φ-ing and S Ψ-ing.

On Craver's account, this condition is trivially satisfied. If a mechanism is *constituted* by its components then a change in any one of them *is* a change in the overall mechanism. Such a coordinated change can be represented as a causal generalization relating changes in the two variables. However, as causal relations do not cross mechanistic levels, on Craver's view, a *Mod-2* generalization is better understood as describing the causal role of x within the overall mechanism S rather than as representing a causal relation between x φ-ing and S Ψ-ing. On this interpretation, *Mod-2* is satisfied just in case x φ-ing can be described independently of other components' activities. This is a weak condition, met by any intelligible mechanistic description of working components. If no activity whatsoever can be attributed to a component, independently of its mechanistic partners, then there seems no reason to distinguish it as a component at all. This is just the counterpart of *Mod-1* for causal activities – a minimal individuation condition for components of MEx.

The distinction between a mechanism's working and effects of its working (§5.5.1) adds another causal dimension. Interlevel causal generalizations do hold, given Craver's assumptions, between components of a mechanism S and a phenomenon P thereby produced. This idea is captured by a third modularity thesis relating x φ-ing and *effects* of S Ψ-ing:

Mod-3: If x is a component of mechanism S, then there is a modular causal generalization relating x φ-ing and P.

Mod-3 requires only that each working component of S makes some causal contribution to P that can be represented independently of the causal contributions of other components to P. Again, this is a plausible requirement. If changes in x φ-ing bring about no change whatsoever in any effect of S Ψ-ing, which could be represented as a causal generalization, then there seems no reason to include x as a component of S. That is, any feature of a mechanism that does not satisfy *Mod-3* is *irrelevant*, for the explanatory purpose at hand. *Mod-3* generalizations are causal and explanatory: a change in x φ-ing produces a change in P. But these modular generalizations explain how P changes, not how S Ψs. Here again tension arises between the multilevel and manipulationist aspects of Craver's account. Together, *Mod-3* generalizations for all the components of a mechanism S yield a fine-grained counterfactual account of how to produce particular outcomes from S. It is not clear, however, that such generalizations amount to an MEx.

A final modularity thesis asserts that they do:

Mod-4: If x is a component of mechanism S, then the MEx of S Ψ-ing (P) includes a modular causal generalization relating x φ-ing to S Ψ-ing (P).

If *Mod-4* is correct, then a MEx is just a collection of modular causal generalizations, arranged in an order corresponding to the mechanism's organization. In realist terms, only modular MEx 'get the causal story right.'[105] If this is the case, then MEx, which aim to correctly describe causal relations among components that together produce an overall phenomenon, should consist of modular causal generalizations. So, if *Mod-4* holds, MEx are causal explanations in the "interventionist" sense (Woodward and Hitchcock 2003). Such explanations reveal what a phenomenon of interest depends on, answering 'what-if-things-had-been-different?' questions about the values of variables within a fixed range of invariance (see §5.4.1). But MEx are prima facie concerned with a different kind of question: 'how-does-it-work?'

This contrast should give us pause. *Mod-4* is a strong modularity thesis, and, unlike *Mod-1-to-3*, is not entailed by the manipulability theory. *Mod-1-to-3* specify necessary conditions for a component of a mechanism (*Mod-1*) that participates in causal relations within that mechanism (*Mod-2*) and causally contributes to its overall effects (*Mod-3*). However, these three modularity theses do not impose sufficient constraints on the component-level to account for MEx. *Mod-4* does impose such a constraint, but holds only if the modular causal generalizations that *in principle* individuate components of a mechanism and their causal roles in relation to effects – the ways we conceptualize them as distinct, relevant components – *are the very same* causal generalizations that figure in MEx. There is reason to doubt this assumption, for MEx in experimental biology. For physiological, cellular, and molecular mechanisms, as we understand them, the behavior of isolated components is not a good guide to their behavior together, and their behavior in one context is not a good guide to their behavior in others. If this line of reasoning is correct, then the causal-mechanical account of MEx leaves out something important: *inter*dependencies among components of biological mechanisms.

5.5.4   Motivating a new account

Together, these three problems undermine the causal-mechanical account of MEx. Though satisfying as an account of how mechanistic models are constructed and tested in experimental fields, it leaves questions about the explanandum, interlevel relation, and representation of components – all the major parts of MEx – unresolved. The target of explanation is ambiguous, the bottom-up direction of MEx elided, and modularity constraints that overcome these obstacles are at odds with what we know of biology. It is important to note that none of these problems are objections to the manipulability theory, which does not (on my interpretation) entail *Mod-4* and is not offered as an account of MEx in biology. *Mod-1-to-3*, which are entailed by Woodward's theory, offer plausible constraints on explanatory descriptions of biological mechanisms. But they do not account for MEx. *Mod-4* does offer such an account, but is not well-supported for experimental biology (see §5.6).

A related concern is that accounts such as Craver's, committed to realism about causes, efface the distinction between mechanisms and MEx. If MEx simply reflect real mechanisms operating in nature, then we need only represent a working mechanism to explain phenomena produced thereby. It is but a short step to the idea that MEx involves pointing out a working mechanism. Mechanism discovery and explanation are then easily run together. But the examples philosophers use to define real-world mechanisms are *MEx*: models of causal mechanisms constructed by scientists. If the aim of MEx is to correspond to real causal mechanisms, and examples of the latter are MEx proposed by scientists, then norms for MEx grounded on such examples can be no more than uncritical reports of scientists' practices. Philosophers would then be limited to describing discovery processes that culminate in MEx. Though the results are likely to please scientists, philosophical contributions to experimental biology should go beyond such reports.

To sum up the analysis so far: prominent law-based and causal accounts of MEx in experimental biology show complementary strengths and weaknesses. The former captures the bottom-up direction of MEx, in that lower levels explain higher ones, yet focuses on physicochemical laws that are peripheral to biological practice. The causal account remains close to biological practice, highlighting the importance of both causal relations and multilevel structure, yet fails to explicate the latter. The next section takes these criticisms in a positive direction.

5.6   Jointness

The core concept of this new account is *jointness*.[106] This term refers to the feature excluded by *Mod-4*: interdependence among causally-active components of a mechanism. Such interdependencies are prominent features of MEx in experimental biology: RNA polymerase binds a DNA promoter sequence to initiate RNA synthesis; multiple testes cells cluster together to form a signaling hub that determines the plane of GSC division; a lymphocyte encountering a pathogen initiates an immune response; and so on. Descriptions of these and myriad other biological mechanisms, offered as explanations of higher-level phenomena, are rife with causally-significant interactions involving two or more components. In this section, I argue that jointly-acting components display the basic structure of MEx in experimental biology.

A classic example of jointness is the 'lock and key' model of enzyme action. This model describes the formation of a complex made up of two components: enzyme and substrate.[107] The enzyme–substrate complex plays a causal role distinct from either component alone, catalyzing a reaction that would not otherwise occur at a physiologically significant rate. Crucially, this causal role is played by the components together, i.e. *jointly*. Enzyme catalysis depends on 'lock-and-key fit' between molecular components of this simple biochemical mechanism. The enzyme–substrate complex, held together by weak chemical bonds, forms in virtue of components' complementary shapes and biochemical properties. The lock-and-key analogy refers to these properties. Causally-active complexes of this kind, and description of the properties enabling lock-and-key fit, are ubiquitous in MEx of biochemical pathways, antigen–antibody binding, cell signaling pathways, the core processes of molecular genetics (DNA replication, transcription, and translation), and many other biological phenomena. Complexes in the GSC example (§5.2) include the hub (an association of cells), the substrate of Jak-STAT signaling (a hub cell, adhesion molecule, and a GSC), the APC (a set of proteins), and the various DNA–RNA–protein complexes involved in chromosome duplication and cell division. Examples can easily be multiplied.

The prevalence of jointly-acting complexes in MEx of experimental biology suggests that such complexes have explanatory significance. I propose, further, that descriptions of jointly-acting complexes, their formation and dissociation, and the properties enabling lock-and-key fit are the 'building blocks' of MEx in experimental biology. In the examples above, jointness is associated with causal activity. Causality is here understood in terms of Woodward's manipulability theory. Jointness is a different relation, holding between causal partners that together produce a particular effect. The issue is *not* number of causes; Woodward's theory straightforwardly generalizes to cases with more than one cause per effect. Jointly-produced effects can be represented as a value (or range of values) taken by an effect-variable. But jointness is not a property of the effect which could be represented by some such value. Instead, jointness concerns the causal role played by a complex of components; that is, ϕ-ing. The difference is reflected in the relevant counterfactuals for each concept. For causality (given Woodward's theory) the relevant counterfactuals concern the relation between values of variables representing causal factors and variables representing effects of those causes. For jointness the relevant counterfactuals compare effects of multiple causal factors separately and together.[108]

The idea of jointly-acting components can be further clarified in terms of complexes:

(J1) Components $x_1...x_n$ jointly ϕ, where ϕ is some causal activity if and only if the complex $x_1...x_n$ ϕ's and uncomplexed x's do not ϕ.

'Uncomplexed x's' include each component alone, as well as multiple x's that do not form a complex. The notion of a causally-active complex sets up a useful distinction. The causal role of complexes within a mechanism requires (i) that each component contributes in some way to the overall mechanism and (ii) that these contributions 'fit' with one another. Requirement (i) is specified by modularity conditions (§5.5.3). The second remains to be clarified. Here the notion of 'mesh,' as with gears of a machine, is helpful:

(J2) Components $x_1...x_n$ form a complex if and only if $x_1...x_n$ are linked in virtue of meshing properties.

Meshing, or complementary, properties enable lock-and-key fit among components – the prerequisite for complex-formation.

Putting (J1) and (J2) together affords a rough definition of jointness:[109]

(J3) Components x_1 and x_2 jointly ϕ if and only if (i) x_1 has properties that mesh with those of x_2 and vice versa; (ii) x_1 and x_2 form a complex x_1x_2 in virtue of their meshing properties; (iii) complex x_1x_2 ϕ's; and (iv) neither x_1 nor x_2 ϕ's individually.

So defined, jointness has three significant features. First, it depends on properties of individual components x_1 and x_2. A complex x_1x_2 forms in virtue of these properties, provided that they mesh. Second, meshing is a relation among just those components; no exogenous initiating activity is required. This contrasts with representation of the action of interlocking gears (the exemplar of meshing parts) as a linear causal chain. Such a chain begins with a 'push' from some outside source. The interlocking arrangement of gears then transmits the productive activity through the rest of the mechanism.[110] For jointness, in contrast, the distinctive causal role of a complex x_1x_2 depends not on some incoming antecedent cause, but on the complementarity of components. The causal activity, ϕ-ing, depends on a particular kind of interaction between x_1 and x_2. In this sense, joint ϕ-ing 'arises from within' out of interacting components. Third, joint ϕ-ing requires a permissive environment, a context in which the complex x_1x_2 can actually form. The most obvious requirement, at least in biological mechanisms, is spatio-temporal proximity such that components interact with one another. But any particular case involves many other background requirements.

In summary, joint causal activity is bottom-up, interactive, and presupposes a context that enables complex-formation. These three features are characteristic of complexes within biological mechanisms, as currently understood. More precisely, causal descriptions of component-complexes within a mechanism satisfy a 'jointness' condition:

(J4) Components x_1 and x_2 jointly ϕ in mechanism S if and only if (J3), and $x_1x_2\phi$-ing is partly constitutive of S Ψ-ing.

This condition is the starting-point for a new account of MEx: the joint account.

5.7   Joint account of mechanistic explanation

5.7.1   Jointness condition for mechanisms

MEx that satisfy (J4) describe individual components and their activities, the properties that allow components to mesh and form complexes, spatio-temporal arrangements that determine what complexes form, causal activities of complexes, and (often) dissociation of complexes. But the requirement that a jointly-acting complex 'partly constitute' the overall working mechanism needs more clarification. To partly constitute a mechanism is to be included in the wider complex that constitutes that mechanism. This shifts the focus, from jointly-acting complexes within a mechanism, to the mechanism *itself* as a jointly-acting complex. The latter notion is entirely compatible with the consensus view: an overall mechanism S is a causally-active (Ψ-ing) complex of interacting parts (ϕ-ing x's). The strengths of Craver's account are retained, with causal relations between productive x's and their effects understood in terms of manipulability. However, the joint account improves on these predecessors by providing a positive account of mechanistic organization from the bottom-up. Much of the organization described in MEx consists of formation and dissociation of complexes of working entities, at particular times and spatial locations within the overall mechanism. The importance of organization among components of MEx is widely recognized. But philosophical treatments of the concept focus on the difference made by varying patterns of causal relations among components (ϕ-ing x's) to the overall working mechanism (S Ψ-ing). Jointness concerns another aspect of mechanistic organization: how heterogeneous components act *with* one another. The three features of jointness, applied to an overall mechanism, explicate the basic structure of MEx: bottom-up and interactive, within a permissive context or environment.

A mechanism's working (its overall behavior) can be defined as the joint activity of its interacting components. So the basic ideas of (M) above (§5.2) can be reframed as a jointness condition:

(JM) Components $x_1,... x_n$ jointly Ψ as mechanism S if and only if (i) $x_1,... x_n$ form causally-active complexes (ϕ_1-ing, ... ϕ_m-ing) in virtue of meshing properties; (ii) these entities and activities are organized so as to constitute S Ψ-ing; and (iii) $x_1,... x_n$ do not Ψ individually.

Condition (JM) explicates the key features of mechanisms: part-whole hierarchy, interactive organization, and an overall context fixed by boundaries of S. This 'joint account' is compatible with Craver's causal-mechanical view, including its norms of accuracy and causal relevance, provided that the latter is understood as an account of MEx *construction and testing* in experimental biology. On the joint account, MEx are not explicated in terms of causal dependency, but constitutive *inter*dependency. More precisely (JM) articulates the basic criterion for MEx: show that a mechanism of interest satisfies this condition, by describing how its components interact to jointly constitute the overall working mechanism. A good MEx shows how working components fit together into complexes, by describing the features in virtue of which they mesh and so make up a complex whole with a distinct causal role. So although description of causal relations is important, the crux of MEx is representation of the constitution relation.

5.7.2 Advantages

The joint account, summarized in (JM), avoids all three problems discussed above. The 'collaborative' aspect of mechanisms in experimental

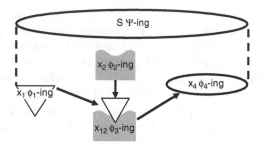

Figure 5.4   The joint account of MEx in experimental biology

biology, elided by *Mod-4*, is the central feature of MEx in these fields. Because jointness depends on meshing properties of diverse components, mechanistic descriptions that conform to (JM) proceed from the bottom-up. The joint account thus captures the directionality of MEx. Finally, it resolves ambiguity concerning the target of explanation, decisively identifying the latter as the overall mechanism (S Ψ-ing), rather than its downstream effects (P). The explanandum is a description of S Ψ-ing, and the explanans a description of components $x_{1...}x_n \phi_{1...}\phi_m$-ing organized to constitute S Ψ-ing. In experimental biology, the latter description includes jointly causal complexes, as well as requirements for complex formation: meshing properties and particular spatio-temporal arrangements of components (Figure 5.4). On this view, a MEx is not a causal explanation of P, but a constitutive explanation of S Ψ-ing in terms of S's working parts.

Importantly, the joint account does not elide the causal aspect of mechanisms, but, rather, places this aspect in proper context. To see this, consider again the multiple levels of causal relations relevant to MEx (Figure 5.3). At the higher level, an overall working mechanism (M Ψ-ing) causes phenomenon P. Such causal generalizations are not MEx, however, but potential explananda for MEx. At the component level, material parts (x's) play causal roles (ϕ-ing) such that each xϕ-ing has some effect on other components. As many of these effects involve formation of complexes, meshing properties, as well as the distinct causal roles of complexes, are included. The component level for a given MEx, therefore, can itself involve multiple levels of complexes and meshing components. The overall mechanism is just the most inclusive complex of a given set of components. Any given MEx describes all these working components and how they are organized to constitute the overall mechanism as a whole. But this description, though it includes causal relations, does not compete with the higher-level causal claim, 'S Ψ-ing causes P.' The two are not alternative causal explanations of P. Biological practice bears this out. For example, molecular and biochemical details of GSC and Jak-STAT do not trump or replace stem cell activities as causes of testis development. Instead, the higher-level claim identifies a particular mechanism as a cause of P. Description of its working parts indicates *how* the mechanism brings P about. The joint account, summarized in (JM), fleshes out the idea that MEx answer 'how-does-it-work?' questions. They do so by placing causal descriptions in context. MEx presuppose, but are distinct from, causal explanations, which aim to answer 'what-if-things-had-been-different?' questions about the values of variables within a particular range of invariance, irrespective of part-whole relations.

The joint account thus explicates key features of MEx, while avoiding the objections to other accounts. The peripherality problem is avoided as there is no appeal to laws or other explanatory devices external to biological practice. The crucial role of organization among diverse components and the constitution relation are both explicated. The joint account does have much in common with Craver's multilevel causal account, but goes further than a friendly amendment in making jointness the crux of MEx.

Jointness also accounts for the fact that MEx in experimental biology typically 'bottom out' at the molecular level. Molecules have a special place in MEx of experimental biology, not because they are governed by physicochemical laws, but because their association into complexes involves all aspects of organization relevant to MEx: spatio-temporal propinquity, change in structural properties, change in causal effects, and, through these, rearrangement of the system as a whole. Complexes of (macro)molecules serve as 'nodes' of mechanistic organization. Description of these nodes seals explanatory gaps in every way that matters for MEx. It is these organizational nodes, not laws governing the behavior of molecules, that serve as the lower bound for MEx in experimental biology. This clarifies the vexed relation between biological MEx and physicochemical laws. The properties required for molecular mesh are themselves explained by physicochemical laws. In this sense, MEx in biology are supported by laws of physics and chemistry. So Weber's account (§5.3) is partly vindicated. But MEx do not thereby reduce to law-based explanations. The two modes of explanation are different. Physico-chemical laws provide solid ground for MEx of biological phenomena, but MEx stand on their own.

5.7.3   Jointness and unification

Finally, the joint account sheds light on the original question: what makes a mechanistic description explanatory? MEx answer how-questions about some S Ψ-ing, where S is a complex system of causally-interdependent parts. These explanations exhibit the constitution relation by describing causal systems at two levels, aligning them such that higher- and lower-level descriptions are represented as identical. When presented with such an explanation, one can see (literally, when the MEx is in diagrammatic form) that an overall mechanism is just its organized, jointly-acting components and vice versa. This identification is a form of *unification*: MEx unify causal descriptions at different levels, bridging the gap between part and whole.

It is important to distinguish this view from explanatory unification in Kitcher's sense (1981). Kitcher's account of explanatory unification shares with the covering-law theory the idea of explanation as logical derivation. On his view, unifying explanations use a single argument pattern to derive descriptions of many diverse phenomena. So explanatory unification for Kitcher is *by* an argument pattern, *of* a wide array of phenomena. Unifying explanations of this sort are simplifying devices, which systematize 'brute facts' into a scientific worldview. MEx also have a systematizing role, but not one that hinges on derivational argument. Their mode of unification instead lends itself to diagrammatic representation. Furthermore, MEx do not typically cover a wide domain of phenomena. The explanatory connection between mechanistic levels concerns the mechanism of interest, whether the latter is a feature of all living things or restricted to only a few artificial contexts. Finally, within their domain, broad or narrow, MEx *connect* rather than simplify. What they connect are not facts per se, but causal descriptions. MEx systematize causal descriptions in a hierarchical structure, with multiple levels defined by the constitution relation. The term 'causal structure' usually refers to a model consisting only of causal relations – a 'pure' causal explanation. MEx are, in contrast, 'mixed,' with causal and constitutive aspects. They help us understand causal relations in the world, by modeling them as hierarchies of complexes. Though not causal structures, strictly speaking, MEx do structure causes.

MEx can, therefore, be improved on at least two dimensions: breadth and depth. Explanatory depth increases with the number of levels, explanatory breadth with expansion of a mechanism's boundaries, bringing additional components and causal relations into the multilevel structure. The joint account thus makes sense of the trend in experimental biology toward increased numbers of mechanistic levels. The more interlevel connections spanned by MEx, the more causal descriptions are unified and systematized, bringing together dependency relations among organisms, organs, cells, molecules, and so on. Here, the contrast with the interventionist account of causal explanation is instructive (Hitchcock and Woodward 2003). Interventionist explanations are "deeper" when invariance under changes in background conditions is "transformed" to invariance under interventions on variables that figure in the dependency relationship at issue (ibid, 188). That is, more of what the phenomenon of interest depends on is made explicit in the explanation. MEx reveal not sources of dependence, but features that allow diverse components to work together, given appropriate background conditions. That is, MEx are not improved by increasing the range of invariance under intervention of causal relations among components of a mechanism. Rather, they are improved by *complementarity* among ranges of invariance, such that complexes become part of one another's environment, transforming background conditions into parts of the mechanism. Like more traditional unifying explanations, MEx increase understanding by revealing connections among many diverse parts. But these connections are not simple, and cannot be concisely stated. Unlike law-based explanations, MEx are built up from details of meshing components and diverse causal relations. Descriptions of these details reveal how diverse parts are unified to jointly constitute an overall working mechanism.

5.8 Conclusion

This chapter builds on recent work in philosophy of biology to clarify mechanistic explanation, the principal mode of explanation for stem cell phenomena. There is strong consensus among philosophers of biology that MEx describe mechanisms composed of diverse working parts (ϕ-ing x's), spatially, temporally, and causally organized so as to constitute the overall working mechanism (S Ψ-ing), which may, in turn, bring about some phenomenon of interest (P). Building on this consensus view, the joint account illuminates key features of mechanisms and MEx, including hierarchical structure, causal interdependence, bottom-up direction, and the significance of the molecular level. This account of MEx compares favorably with the main alternatives, which invoke laws and causal relations, respectively, as loci of explanatory power. Though the manipulability theory of causality fruitfully extends the consensus view of mechanisms, it does not fully account for MEx in experimental biology. MEx do not aim to identify causes, but to show

how those causes work within a framework of part-whole hierarchy. Modularity theses that are plausible for biology do not fully account for MEx, while jointness explicates its basic structure and source of explanatory power.

The explication provided here is, like MEx in stem cell biology, partial and incomplete. Even in this rough form, however, the joint account improves our philosophical accounts of MEx. It highlights a hitherto neglected aspect of organization in biological mechanisms, parts working *together*, and reveals the link between this prominent feature of biological mechanisms and the hierarchical structure of MEx, thereby reconciling bottom-up and multilevel aspects of these explanations. Most importantly for my purposes here, the joint account has significant implications for stem cell biology, particularly for clarifying concepts of stemness and cell state. As discussed in Chapter 4, stemness and cell state should be understood as hierarchical explanatory sketches, encompassing both cellular and molecular levels. On this view, a cell is conceived as an overall system (S) that exhibits some behavior of interest (Ψ-ing): engaging in developmental processes, functioning within an organism or tissue, interacting with other cells, and so on. Some behaviors, including self-renewal and differentiation, produce further effects in turn: new cells, cell populations with different traits, and more mature organisms (P). How a cell Ψ's (so as to produce P) is explained by organized molecular components (x's), which interact (jointly ϕ) to constitute the cell state. Underlying molecular mechanisms therefore explain an overall cell state. This idea is explored in later chapters. However, I next consider another implication of the joint account: all working components described in a given MEx are equally contributors to the explanation of the higher-level phenomenon of interest. So MEx do not privilege a single 'master cause.' This contradicts the claim, often made by biologists and philosophers alike, that genes have a privileged role in explanations of biological phenomena. The next chapter examines the contested role of genes in development.

6
Genes and Development: The Stem Cell Perspective

6.1 Introduction

The idea that genes control development is widespread in biology and philosophy. But the notorious ambiguity of the term "gene" makes assessment of this claim difficult.[111] In classical genetics, "the gene" is a theoretical entity that predicts patterns of traits across generations. Localization of this hereditary material to chromosomal DNA and elucidation of its structure in the 1950s cemented the more realistic molecular gene concept. The Central Dogma (symbolized as DNA→ RNA→protein) and genetic code then established molecular genes as the primary causes of phenotypes. Today, a common understanding of the term is:

(G) A gene is a causally active DNA sequence associated with a particular phenotypic trait.

This formulation leaves the nature of the association between genes and traits open. If it is causal, then we have:

(G') A gene is a DNA sequence that causes a particular phenotypic trait.[112]

On this "causal-informational" view, a gene embodies the plan or program for a specific trait. It is then but a short step to the view that development is just the execution of a pre-determined genetic program. Twentieth-century triumphs in genetics have bolstered the view that genes are causal agents of distinctive significance for organismal development and resulting phenotypes.

However, this "genetic exceptionalism" is vulnerable to two serious objections. The first is "Lillie's paradox:"[113]

> It is... almost universally accepted genetic doctrine today that each cell receives the entire complex of genes. It would therefore appear to be self-contradictory to attempt to explain [differentiation] by behavior of the genes which are ex hyp. the same in every cell... The essential problem of development is precisely that differentiation in relation to space and time within the life-history of the individual which genetics appears implicitly to ignore... Those who desire to make genetics the basis of physiology of development will have to explain how an unchanging complex can direct the course of an ordered developmental stream.
>
> (Lillie 1927, 365–367)

The problem can hardly be put more starkly. All cells of a multicellular organism (barring mutations and other rare exceptions) share the same DNA sequences. But development involves production of differences among an organism's cells, tissues, and organs. How can invariant genes account for developmental diversity?

The solution is to invoke differences in *gene expression* among cells and tissues.[114] The Central Dogma represents gene expression as a simple causal-informational chain: a sequence of chromosomal DNA is a linear template for a mRNA transcript, which is, in turn, a template for a sequence of amino acids making up a protein. Cell phenotype depends on which genes are transcribed and then translated. But this raises a second objection: DNA does not express itself. Fifty years of molecular biology have revealed a menagerie of mechanisms implicated in the complex cellular machinery of gene expression. Transcription and translation are performed by elaborate protein–RNA complexes. The same chromosomal DNA may yield different proteins, depending on which other genes are expressed in a given cell. Chromosomal DNA also includes non-coding regulatory elements that affect gene expression: introns, promoters, termination sequences, upstream and downstream activators and enhancers, and multiple families of repetitive sequences. Non-DNA components of chromosomes, such as histones, modifier groups, and chromatin conformations, also influence gene expression. The path from DNA to protein is further complicated by mechanisms that modify gene products in multiple ways, including proofreading, RNA processing and editing, and alternative splicing. Yet another layer of RNA regulation was discovered in 2001: micro-RNAs, which modulate gene expression by specific sequence binding. Variation in gene products continues post-translation: proteins are spliced, folded, tagged for distribution within or without the cell, and cooperatively bind one another.

The complexity and diversity of gene expression mechanisms makes it impossible to identify any single molecular entity as a context-independent gene (G). What counts as a gene depends on the organismal and cellular context, as well as the trait of interest. Multiple layers of regulation allow for many ways of individuating "the gene," even for a particular protein. This inveterate pluralism vitiates the claim that genes (G) control development. But a new defense of genetic explanatory privilege is available, building on the manipulability theory of causality (see Chapter 5). Its core thesis is that genes, but few other components of molecular mechanisms, are "specific, actual difference-making causes" of development (Waters 2007). I first unpack the terms of this thesis and then assess it in regard to stem cell biology.

6.2   Genes as difference-makers

According to the manipulability theory, causes *are* difference-makers. More precisely, X is a cause of Y just in case different values of a causal variable X make a difference to the value of an effect variable Y, such that this relation is stable under some interventions. So only the terms "specific" and "actual" require clarification here. Waters (2007) defines the latter as follows: X is *the* actual difference-maker with respect to trait Y in population P if and only if X causes Y in Woodward's sense; this relation is invariant with respect to other variables that actually vary in P; the value of Y actually varies among members of P; and actual variation in X fully accounts for the variation in Y in P.[115] These conditions, he argues, mark an "objective difference" between actual and merely potential difference-makers. Waters' account acknowledges the molecular complexity of gene expression in that "DNA is only one of many causes" of biological phenomena and "exercises its roles through the production of RNA and polypeptide molecules" (2007, 552–553). However, not all causes of development are ontologically "on par." Genes, as actual difference-makers, are responsible for phenotypic variations among members of actual populations.

This objective difference among biological causes underwrites an explanatory privilege for genes in development. The effect explained is a difference in the value of a variable Y in some population of developing entities. The explanatory cause is a variable X that can take multiple values and satisfies the conditions for an actual difference-making cause.

Differences in DNA sequence that correlate with different phenotypic traits in a real population meet these criteria. In contrast, molecular and cellular entities that make a causal contribution to development, but do not vary within the population of interest, are merely *potential* difference-makers. Genes (G) are therefore distinctively responsible for actual differences in a population of cells or organisms. This conception of genes dates back to the early twentieth century and T. H. Morgan's research program of classical genetics. Morgan conceived the *difference* in many-to-many relations between genes and traits as the attribute solely *of the gene* (1926, 322). His differential gene concept accords with the manipulability theory. Both shift the focus of explanation from the nature of a cause to what makes a difference to the value of a variable under controlled conditions. Experiments designed to reveal causal relations pick out difference-making causes.[116] Waters' application of the manipulability theory is thus supported by classical genetics.

However, being constrained to accommodate molecular genetics as well, Waters' account limits genetic explanatory privilege in several ways. First, genes are actual difference-makers only relative to a population. This restricts gene-based explanations to a particular genetic and environmental context, within which "uniform phenotypic differences" are caused by differences in DNA (Waters 2007, 553). Second, the explanandum for which genes offer a privileged explanation is a *difference* among members of a population, not development per se. Finally, Waters' argument concerns only the "small part of development" described by the Central Dogma, namely specification of RNA/protein sequence by DNA sequence. So his defense of genetic explanatory privilege is restricted to the "coding" relation between a population of DNA sequences and a population of RNA or protein molecules within a cell.

This difference-making relation has another distinctive feature: specificity. DNA is distinctive, Waters argues, in that different specific changes to its sequence produce different specific changes in molecular products (2007, 574–575). Other components of cellular machinery relate to the latter more simply, like on/off switches. DNA's specificity permits finer modulation of phenotypic states. Woodward (2010) further explicates this concept.[117] A classic illustration is the lock and key model of enzyme action (see §5.6). In this (disconfirmed) model, each enzyme binds a unique substrate to catalyze a unique chemical reaction. A single cause (enzyme) produces one kind of effect (reaction), while each effect is produced by a single cause. Specificity in this "one cause-one effect" sense is relative to a pre-specified class of alternative causes and a pre-specified range of kinds of effect. Because there are no fixed criteria for distinguishing among alternative causes and bounding the ranges of effects, this concept remains somewhat vague.

Further clarity is provided by a second specificity concept, defined in terms of a mapping from states of a cause X to states of an effect Y (Woodward 2010, 305). The more closely this mapping approximates a bijective function, the more specific the causal relation between X and Y. The key idea here is "proportional influence" (hereafter PI) of causes on effects. If we can vary the state of the cause, then we can determine which state of the effect is realized, out of a range of alternatives. Nonspecific causes are such that many different states of X map to the same state of Y, or the same state of X maps to many states of Y, or both. So, like 'actuality,' PI-specificity provides a basis for discriminating among the various causes of an effect. Relative to sets of alternative values of an effect variable and each of its causal variables, more PI-specific causes afford finer-grained control over the value of the effect. If fine-grained control is advantageous for explanation then there is reason to privilege PI-specific causes in explanations over non-PI-specific causes of the same effect. These two concepts of causal specificity, PI- and one-cause-one-effect, are not precisely equivalent.[118] But, for present purposes, they can be treated together. The explanations at issue are mechanistic explanations (MEx). In constructing these explanations, the overall mechanism, associated phenomena, and components are delineated, thereby fixing the sets of alternative values to be considered. Relative to such a set, one-cause-one-effect and PI-specificity coincide. For simplicity, "PI-specificity" hereafter refers to both concepts.

Augmented with the concept of PI-specificity, Waters' view amounts to a sophisticated form of genetic reductionism. The key idea is that, though MEx describe many causal relations, those involving genes and their effects (RNA and protein sequences) are distinguished by PI-specificity. In this sense, genes are fundamental to MEx of development. Unlike linguistic and information-theoretic treatments, PI-specificity casts DNA as a fundamental template on the basis of causal relations relevant to experiment. This is a promising idea. The next two sections critically examine this sophisticated gene-centric view.

6.3 Specificity and mechanistic explanation

PI-specificity bears on MEx in several ways. First, PI-specific causal relations are good starting-points for these explanations. This is because PI-specific input–output relations indicate the boundaries of a mechanism of interest, while conditions that reveal PI-specificity often indicate key components. Second, PI-specificity guides construction of MEx, identifying relevant components and the appropriate grain of description for their workings. Third, PI-specificity motivates MEx in practical contexts. Biotechnology, for example, often exploits PI-specific causal relations in mechanisms that we construct. But to intelligently use a PI-specific causal relation we need to know how its mapping is achieved. If we know how a fine-grained mapping of input and output conditions is brought about in one context, we can make inferences about its stability in other contexts, as well as possible unintended consequences. Furthermore, as mechanisms are often components of other mechanisms, causal relations within mechanisms we engineer are often PI-specific.

These points notwithstanding, PI-specific causal relations are neither necessary nor sufficient for MEx. That they are insufficient is straightforward. The enzyme RNA polymerase I (hereafter RNA pol I) is a crucial component of the mechanism by which a DNA sequence is transcribed to produce a mRNA. Many different states of this protein are possible, which vary in sequence and shape. But states of RNA pol I do not map onto states of transcribed mRNA so as to approximate a bijective function. Many different states of RNA pol I are compatible with its functional role and, from each of these, many mRNA sequences can be produced. Many other states are non-functional and all map to the same effect: no mRNA. So RNA pol I is non-PI-specific. Yet no adequate MEx of transcription could omit this protein. Description of actual PI-specific difference-making causes is not sufficient.

Nor is PI-specificity necessary for MEx. Consider the lock-and-key model again. In mechanistic terms, the lock and key relation holds not between enzyme and chemical reaction, as in PI-specificity, but between enzyme and substrate. Binding of these two components into a complex catalyzes one or more chemical reactions. Many experiments have shown that the causal relation between enzymes and reactions is *not* PI-specific. But enzymatic reactions are well understood mechanistically. Enzymes and substrates bind specifically to one another in virtue of certain molecular features. As discussed in Chapter 5, binding 'partners' have complementary shapes and biochemical properties that, given certain spatio-temporal conditions, cause them to form a complex via weak chemical bonds. Enzyme–substrate complexes play a distinct causal role, which the same components, dissociated, cannot. Lack of PI-specificity does not undermine MEx of enzyme catalysis.

It is even possible that PI-specific causal relations are the exception rather than the rule in biological mechanisms. Redundancy, multiple interactions per component, and mutual adjustment of parts are all characteristic of developmental mechanisms, and all involve deviations from PI-specificity. MEx of developmental phenomena remain incomplete, despite enormous progress in identifying genes and PI-specific molecular relations, precisely because of the prevalence of non-PI-specific relations in development. A different concept of specificity, concerning organization of interactions among heterogeneous components, may be more relevant for MEx. As the previous chapter showed, the overall working of a mechanism typically depends on exactly which components interact with one another, as well as when, where, and in what context. Mechanistic descriptions must be specific in this respect. But this is distinct from PI-specificity. Hereafter, the term "specificity" refers to the interactive, organizational conception.

Though description of PI-specific causes is neither necessary nor sufficient for MEx, such explanations do describe actual difference-makers. MEx aim to describe how some mechanism actually works, or, sometimes, how a phenomenon P is actually brought about. Experiments that reveal the answers show that manipulating a component makes a difference to the overall mechanism and to interactive processes within it. So both an overall mechanism and its components are actual difference-makers. But the crux of MEx is not identifying difference-makers as such. Rather, a MEx shows how interacting components jointly constitute the overall mechanism, unifying levels of biological organization in a particular context (see §5.6). MEx do not show *that* some mapping from cause to effect exists, but describe *how* the mapping is made. Within a MEx there is explanatory parity; all components are equally implicated. So the actual/potential and PI-specific/non-PI-specific distinctions do not underwrite a privileged role for genes (G) in MEx of development, though they may do so in other kinds of explanation.

6.4   Cell reprogramming

I next consider a second argument for genetic privilege, based on cell reprogramming experiments. Reprogramming experiments manipulate cell development by adding specific DNA sequences to cultured cells. The first reprogramming experiments used four genes (Oct3/4, Sox2, Klf4, and c-Myc) and mouse skin cells (see §2.5.1). The genes were "delivered" to cells using engineered retroviruses, which infected cells and integrated new DNA sequences into their chromosomes. A very small percentage (~0.05%) of cultured cells then transformed their morphology, gene expression, and developmental capacities, becoming similar to embryonic stem cells (ESC). These 'reprogrammed' cells can self-renew indefinitely or give rise to many different cell types. They are accordingly termed 'induced pluripotent stem cells' (iPSC). Reprogramming evidently makes a difference among cells, and it is reasonable to suppose that DNA sequences are the actual difference-makers in this case. But, surprisingly, they are not. The actual difference-makers in reprogramming experiments are complex molecular mechanisms that include, but are not controlled by, particular DNA sequences.

The first hint that more is involved than difference-making DNA is that reprogramming with genes is "a slow and inefficient process consisting of largely unknown events" (Maherali and Hochedlinger 2008, 595). At best, only a small percentage (<0.1%) of gene-treated cells become iPSC. Difference-making relations do hold at the population level. But the striking inefficiency of reprogramming for individual cells highlights our lack of understanding of how the process works; that is, the mechanism by which cells are transformed. DNA components of the mechanism, the four reprogramming factors, are well characterized and can be precisely controlled. But these genes are not sufficient. Experiments aimed at revealing reprogramming mechanisms therefore target other components and interactions as well: levels and timing of gene expression within cells, biochemical interactions that affect chromosome structure, intercellular signaling pathways, and more. Since 2006, the original iPSC method has been refined and modified by hundreds of different laboratories. The result is not a single standardized procedure, but a multistage schema indicating variables that make a difference to experimental outcome (Figure 6.1).[119] Five kinds of variable are distinguished: (i) reprogramming factors, (ii) delivery method, (iii) starting cell population, (iv) factor expression, and (v) culture conditions. Different combinations of values of these five variables are more or less effective at producing iPSC with desirable developmental capacities, such as the ability to produce neurons, blood, or cardiac muscle in vitro.[120]

For any particular experiment, values of these variables are selected according to researchers' purposes and mutually optimized, insofar as current knowledge permits. If reprogramming is conceived as a causal process that makes a difference to cells in the starting population (iii), then variables (i), (ii), (iv), and (v) are all actual difference-makers. Moreover, DNA is not required for (i). RNA and protein sequences can also reprogram cells to iPSC. So progress in reprogramming since 2006 does not single out DNA sequences as the actual difference-makers. Instead, the emerging picture is of diverse molecular mechanisms involving multiple interacting variables.

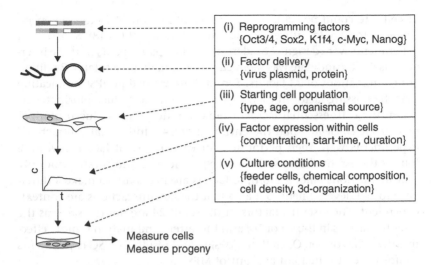

Figure 6.1 General scheme of iPSC reprogramming

Furthermore, genes were not actual difference-makers in the original 2006 experiments that first yielded iPSC. Researchers at Kyoto University began with the hypothesis that "the factors that play important roles in the maintenance of ES cell identity also play pivotal roles in the induction of pluripotency of somatic cells" (Takahashi and Yamanaka 2006, 663). They used this hypothesis to construct a list of 24 candidate factors, including genes *and* proteins. Genes highly expressed in tumors and ESC were selected, as were proteins implicated in pluripotency mechanisms in embryos and ESC.[121] Genes, as such, were not given priority as candidates to induce pluripotency. Instead, all three molecular forms in the Central Dogma – DNA, RNA, and protein – were subsumed under the inclusive term "factor." Next, the 24 candidates were whittled down to a "core set:" Oct3/4, Sox2, Klf4, and c-Myc. Each member of the core set was characterized both as DNA and as protein (Yamanaka 2007, 43–45). So throughout the experiment proteins were as significant as DNA. Difference-making factors were individuated and tested in ways that did not discriminate among these molecular forms.

In addition, the method by which the core set was generated is designed to reveal jointness rather than PI-specificity. First, the Kyoto team showed that all 24 candidates together are sufficient to induce pluripotency. Next, they added each factor to cells individually, which yielded no iPSC. This demonstrated that interactions among factors are necessary to make a difference to cell development; "drastic alterations of cell fate could be achieved with a combination of factors when no single factor would suffice" (Cohen and Melton 2011, 248). Subsequent experiments were designed to identify the key participants in these interactions. Each candidate was *subtracted* individually for a total of 24 different combinations of 23 factors. Ten of these yielded no iPSC, indicating that the missing factor in each case is necessary to induce pluripotency. When these 10 essential factors were combined, they induced pluripotency *more efficiently* than the entire set of 24. "Individual subtraction" experiments were then performed for each essential factor, revealing four as the essential core set. These experiments demonstrated not only that interactions among multiple factors are necessary to make a difference to cell development, but also that the key interactions are context-dependent. The essential factors in the set of 24 are not the same as the essential factors in the set of 10, and the latter combination is more effective than the former. Overall, the design and results of iPSC experiments conform well to the joint account of MEx.

Although current MEx of reprogramming are works-in-progress, some robust features have already emerged. Most strikingly, all members of the original core set, and all comparably effective alternatives, are *transcription factors* (TF). As proteins, TF bind specifically (in the mechanistic sense) to particular sequences of DNA, which, in turn, affects transcription of genes nearby on the chromosome (Figure 6.2). TF-binding DNA sequences are regulatory rather than protein-encoding; the mechanism of transcription involves both. These regulatory sequences are distributed non-randomly in the genome; notably, the same sequence often appears near genes with related functions. So altering the expression of one TF gene can make a difference to the expression of dozens, even hundreds, of other genes – including other TF genes. TF proteins are thus "phenotypic switches" that can coordinate large-scale patterns of gene expression within a cell.[122] The crux of coordinated switching is the binding interaction of a TF protein and a regulatory DNA sequence. Neither component has priority over the other; both DNA and protein are crucial. In emerging MEx of cell reprogramming, DNA sequences play a significant role – but not a controlling one. Genes are interactive partners with other components, rather than "masters" dominating minions.

Reprogrammers' terminology is also inconsistent with the idea that genes are privileged components of developmental mechanisms. In experimental biology, by convention, names of genes are written in italics, names of proteins in regular type (for example, *Nanog* and Nanog).[123] Though stem cell biologists typically respect this convention, discussions of reprogramming often shift rapidly between DNA and protein terms,

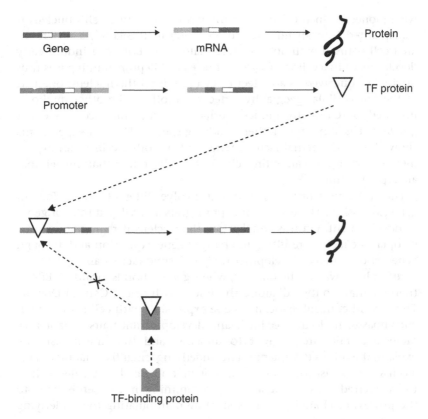

Figure 6.2 Role of transcription factors

or use regular type only.[124] Terms such as 'Oct3/4' may refer to that TF's DNA sequence, RNA transcript, or functional protein. Indeed, the inclusive term 'factor' is plausibly taken as referring to all three, given that the mechanistic role of TF is to mediate between DNA, RNA, and protein. On this interpretation, 'Nanog,' for example, refers to an entire TF unit: a compressed version of the Central Dogma.

Increased understanding of TF mechanisms has led to a shift in the concept of reprogramming itself. Intuitively, the term suggests a pre-existing plan for cell development, which experimenters *re*program by adding exogenous TF. But informational metaphors are neither prevalent nor powerful in stem cell biology today (Brandt 2010). "Cell reprogramming" refers generally to experimental methods that induce pluripotency. There are four such methods; iPSC production by TF being the newest. Two of the methods, nuclear transplantation and cell fusion, were pioneered in the 1950s. In the former, a mature cell's nucleus is transplanted into an undifferentiated egg; in the latter, a differentiated cell is fused with an undifferentiated cell. Both can induce early development in a cell with an 'old' nucleus.[125] Reprogramming was thus originally conceived as an effect of cytoplasm on the nucleus; one part of the 'intracellular geography' affecting another. The other methods, invented more recently, induce pluripotency in cultured cells by either growing cells atop one another or adding specific TF. Such experiments show effects of external factors, whether from other cells or an experimenter's syringe, on an entire cell. So it is cells, rather than nuclei, that are reprogrammed.[126]

The two senses of reprogramming involve different conceptions of gene expression. The earlier concept suggests a privileged role for genes in development. On this conception, the nucleus is reprogrammed by cytoplasmic factors, resulting in changed gene expression and, through gene action, altered development. Cytoplasmic factors affect development only by way of the nucleus, where gene action is localized. Effects then emanate 'outward' along the linear path of the Central Dogma. The second concept associates gene expression with cell state, which encompasses molecular, cellular, and developmental traits (Chapter 4). Molecular traits are thought to underlie, and thus mechanistically explain, the others (Chapter 5). The underlying 'circuitry' that maintains a cell state consists of DNA, RNA, protein, and small molecules, collectively referred to as a 'program.' *Re*programming, in this sense, refers to the process of changing a cell's state by manipulating the underlying molecular program.

TF play a distinctive role in these programs, acting as 'switches' for manipulating cell state. TF can therefore qualify as actual difference-makers for cell development. But they are not PI-specific. The relation between TF and cell state is many-to-many. Moreover, TFs do not act alone, but jointly. Explanations of differential developmental outcomes go beyond identification of TF, to describe entire mechanisms. For example, three TF (Oct4, Sox2, and Nanog) are proposed as a "core pluripotency network" that can act as a switch for developmental potential in mouse and human cells. This network is thought to play two key roles: repress genes associated with differentiation, and activate ESC-specific genes (Stadtfeld and Hochedlinger 2010, 2249). Preliminary MEx of both activation and repression describe binding of protein complexes to "cognate DNA sequences." DNA sequences are not privileged in these explanations. Instead, emphasis is on specific interactions among components, that is jointness.

To sum up: the concepts of reprogramming, cell state, and mechanistic explanation are closely connected in stem cell biology today. Reprogramming is increasingly conceived as change in cell state, though the earlier sense of the term also persists. Cell state implicates both molecular and cellular levels, giving the new sense of reprogramming a dual character. The two levels correspond to the multilevel structure of MEx for cell development, as explicated by the joint account (Chapter 5). Mechanisms of cell reprogramming are constituted by diverse components that selectively associate so as to jointly bring about a change in cell state. Within these mechanisms, all components are equally crucial; none is privileged over others. Genes do not play a controlling or foundational role in MEx of cell state changes. The reprogramming case does not support the view that genes play a privileged role in development as actual, specific difference-makers.

6.5 Waddington's epigenetic[127] landscape

A better representation of genes in development, with particular relevance for stem cell phenomena, is Waddington's landscape. Conrad Hal Waddington articulated the landscape metaphor for development in several texts, most extensively in *The Strategy of the Genes* (1957).[128] This simple model, originally constructed to unify embryology and genetics, has been recently co-opted by stem cell biologists to represent reprogramming experiments. The landscape's structure, representational assumptions, and relation to experiment, in both original and updated versions, clarify the role of genes in development.

The landscape model is a two-dimensional diagram of a three-dimensional structure (Figure 6.3):

> ...a more or less flat, or rather undulating surface, which is tilted so that points representing later states are lower than those representing earlier ones... Then if something, such as a ball, were placed on the surface, it would run down toward some final end state at the bottom edge.
>
> (Waddington 1957, 29)

The axis projecting outward to the viewer represents time. The horizontal axis represents phenotype, ordered by some measure of similarity. The vertical axis represents the order of development; the surface's tilt correlates this developmental order with time. A rolling ball's path down the incline corresponds to development of some part of an organism

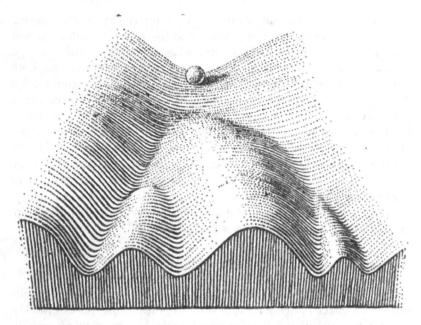

Figure 6.3   Waddington's landscape model (from Waddington 1957, 29)

from an early undifferentiated state to a mature differentiated state. The bottom edge describes a series of dips representing alternative mature states, while the top edge describes a curve with a single minimum, representing the undifferentiated start of development. The undulations of the landscape carve it into valleys with the form of branching tracks. These valleys connect the initial state with multiple discrete end-states. So the landscape represents developmental potential gradually restricted over time, partitioned into diverging channels. One could say that it diagrams the concept of 'normal development' as a simple geometric structure. For Waddington, the model served three related purposes, which are reflected in its structure and representational assumptions. I discuss each in turn.

6.5.1   Developmental pathways

Most obviously, the landscape model represents patterns of phenotypic change in developing tissue. Prominent valleys represent resistance of these patterns to external perturbation: *robustness*. The developing entity itself is left unspecified;[129] the *pathways* of development are the main representational target. In other words, Waddington's landscape represents developmental *potential* rather than development per se. The ball is poised at the top, with the landscape fanning out below. In this position, the ball represents a fertilized egg or a portion thereof; Waddington does not specify an interpretation for other positions. Indeed, the distinction between surface and ball is somewhat arbitrary. In a 1956 diagram, Waddington identifies the top edge as representing an early developmental stage, with different points representing "intrinsically different" regions of egg cytoplasm (1956, 351). Alternatively, and equivalently, different regions of egg cytoplasm could be represented as balls having different initial biases that influence their trajectories. This flexibility makes it difficult to specify the representational target of the ball at intermediate positions, other than generically, as 'the developing entity.' But this interpretative difficulty simply reinforces that the developing entity is not the model's representational target. Rather, Waddington's landscape represents developmental pathways, the pre-determined options available to the fertilized egg or a part thereof.

This representational target can be analyzed further. The landscape model exhibits three "essentially formal" properties of development: unidirectionality in time, multiple discrete termini from a single undifferentiated start, and robust bifurcating tracks (1957, 49). These structural features of the model reflect generalizations about animal development, which are, in turn, based on empirical observations and experiments. Unidirectionality in time and multiple discrete termini from an undifferentiated start-point are features taken as essential to organismal development, supported by observations dating back to Aristotle. Robustness and bifurcating structure of developmental pathways, however, are generalizations grounded in twentieth-century experimental embryology. The model represents two such generalizations: first, that developmental pathways tend to "self-stabilize" in the face of minor perturbations and, second, that certain steps in developmental pathways depend on a (nonspecific) stimulus. Waddington inferred both from the combined results of experiments on *Drosophila*, chick, and various marine organisms. The first generalization is based on experiments that varied the physical or chemical environment of a developing embryo. Results showed that for many such interventions, not only developmental outcomes, but also intermediate stages, remained constant.[130] Waddington concluded that developmental pathways tend to self-stabilize in the face of minor perturbations of internal or external origin. Their robustness is represented in the model by valleys' depth and steepness. A valley with steep walls corresponds to an interval of a developmental process in which the fate of the tissue in question is pre-determined, barring extreme perturbations.

A valley with gently sloping walls corresponds to an interval in which the tissue is responsive to internal or external stimuli, such that small disturbances can 'push' development onto another path.

Other embryological experiments indicated that developmental pathways include intervals of stimulus-dependence. For example, in chick and *Drosophila*, neural ectoderm forms an eye lens only if an inducer is present at a particular time; otherwise, no lens develops. During the crucial interval, diverse substances – other tissues, chemicals, objects introduced by the experimenter, even artificial compounds that would never be encountered in normal development – can induce lens formation. So the inducing substance does not specify the fate of the developing tissue. These results were obtained by experiments that isolated parts of a developing organism and then added or removed certain environmental factors at particular times. To explain the results, Waddington distinguished between a tissue's competence to respond to an inducing stimulus and its potency, or ability to give rise to a range of specific developmental outcomes. The developmental potency of neural ectoderm, at least in some organisms, is determined in advance as a choice of two options. Which of the two is chosen depends on the presence or absence of a stimulus when the tissue is in a competent state. Alternative pathways are represented in the model as bifurcating tracks. Waddington used the landscape analogy to illustrate the distinction between competence and potency for developmental processes in general.[131] So the model's prominent structural features, notably its branching valleys, are grounded on experiments that manipulated development in model organisms.

6.5.2   Genetic control

Waddington's landscape also expresses a hypothesis about the role of genes in development: that genes indirectly control development through a network of interacting biochemical products. The idea is depicted in a companion diagram showing the 'underside' of the landscape, which reverses the viewer's perspective (Figure 6.4). Seen from below, the landscape's surface "slopes down from above one's head towards the distance" (1957, 36). The contours of the landscape are "controlled by the pull of these numerous guy-ropes which are ultimately controlled by genes." Guy-ropes represent gene products, while their webby connections represent biochemical interactions among those gene products. The underside diagram shows gene products organized into interacting networks that directly determine the landscape's topography. The ultimate determinants, however, are genes, depicted as fixed pegs. Their fixity represents genes' lack of alteration during development, and suggests a

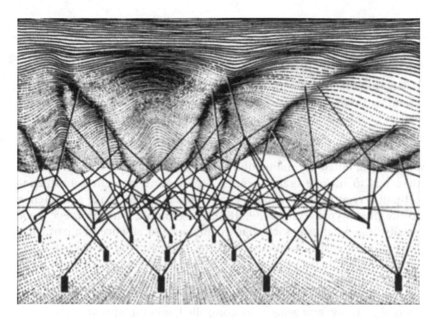

Figure 6.4 Genetic underpinnings of Waddington's landscape (from Waddington 1957, 36)

kind of ultimate control. Waddington's hypothesis about genetic control of development is visualized by the landscape's two sides: a top-side view of branching pathways leading to stable developmental states, and an underside of genes and their interacting products.

The model visually unifies development and genetics via these two complementary images of the landscape: robust pathways above and interacting gene products below, with genes at the bottom metaphorically 'pulling the strings.' Development and genetics are further unified via the branching-track structure, which visualizes a formal analogy between cellular, genetic, and organismal development. Branching tracks have been used to represent cell development since the late nineteenth century, most prominently in cell lineage diagrams that track cell pedigrees and division events.[132] In these models, branch-points represent cell division events and branching tracks represent genealogical relations among cells. Other cell characteristics are also represented, such as position within the developing embryo, morphology, and developmental fate. Waddington's landscape is evidently not a cell lineage diagram, as it does not represent cell properties, intercellular relations, or, indeed, *cells* at all, apart from the fertilized egg (see note 129). The landscape's branch-points represent choices among developmental pathways, not mitotic events. Nonetheless, as Gilbert (1991) persuasively argues, the structural correspondence between cell lineage diagrams and Waddington's landscape draws a formal analogy between cell and organismal development.

The structural correspondence with models of gene action goes deeper. Waddington first articulated the landscape analogy in 1939 as a generalization of time- and dose-effect curves representing the role of genes in producing specific phenotypic effects. In these genetic models the effects of different alleles of a single gene are plotted against time, with "doses" differing for different combinations of alleles. Waddington modified these diagrams to include branch-points, representing steps in a biochemical pathway at which differences in a gene's product make a difference to the phenotypic result (1939, 182). For example, *Drosophila* eye color results from a pathway such that the presence or absence of a particular gene product makes a difference to the pigmentation in the adult fly. Generalizing from data on *Drosophila* mutants, Waddington hypothesized that genetically-controlled switch-points underlie the developmental pathways that produce phenotypic traits. The bifurcations of developmental pathways and genetically-controlled switch-points are thus different aspects of the same process.

'Branching' representations of gene action depict the difference that mutant versus wild-type alleles of a gene make to phenotypes, while branching tracks on the landscape represent the possibilities available to developing tissues. Because multiple genes were usually implicated in a single pathway, such as that producing *Drosophila* eye color, branching-track models naturally expanded to include effects of multiple interacting genes. The landscape model results from including interactive effects of the entire genome on a particular pathway:

> If we want to consider the whole set of reactions concerned in a developmental process such as pigment formation, we therefore have to replace the single time-effect curve by a branching system of lines which symbolizes all the possible ways of development controlled by different genes. Moreover, we have to remember that each branch curve is affected not only by the gene whose branch it is but the whole genotype. *We can include this point if we symbolize the developmental reactions not by branching lines on a plane but by branching valleys on a surface.* The line followed by the process, i.e. the actual time-effect curve, is now the bottom of a valley, and we can think of the sides of the valley as symbolizing all the other genes which cooperate to fix the course of the time-effect curve; some of these genes will belong

to one side of the valley, tending to push the curve in one direction, while others will belong to the other side and will have an antagonistic effect. One might roughly say that all these genes correspond to the geological structure which moulds the form of the valley.

(Waddington 1939, 182, emphasis mine)

6.5.3 Unification

Waddington's branching-track representation of gene action was speculative, conforming to his unificatory aims rather than contemporary standards in genetics (Gilbert 1991). These unificatory aims also extended to evolution. To bring evolution into the picture, Waddington conceived the landscape as malleable, such that changes to a gene or interactions among products alter its topography. These changes represent modification of developmental pathways on evolutionary time-scales. Waddington further speculated that developmental pathways bias genetic change in particular directions, forming a feedback loop of evolutionary and developmental change mediated by genes. He intended the landscape as a "conceptual laboratory" for visualizing theoretical ideas about these interrelated processes of change. One influential example is *canalization*, the process by which developmental pathways become more robust to perturbations. The more canalized a pathway, the more stimuli can induce a tissue to enter it and the less stimuli can turn a tissue from it. Canalization is visualized on the landscape as increased steepness of valley walls. In the model, such topographic changes are produced by changes in the underlying network of interacting genes and products: changes in the placement of pegs or tensions on the guy-ropes. So the model offers an intuitive illustration of evolution of development via genetic change. Malleability of the landscape is crucial to such illustrations, representing interrelated developmental, genetic, and evolutionary processes.

Its role as conceptual laboratory also constrained the landscape model to be simple. Waddington was well aware that a rigorous treatment of development in a geometrical framework would add a dimension for each phenotypic trait, with each point representing a complete state description of the developing entity in multi-dimensional space (1957, 26, 49). Yet he constructed the model in three dimensions. Phenotypes are represented on a two-dimensional surface, one dimension of which is correlated with time. So features of the model that could be used to derive predictions (such as steepness of valley floor or walls compared with the overall slope of the landscape) lack a principled theoretical interpretation. The rate of developmental change over time or the degree of a pathway's robustness, for example, can be represented only qualitatively, on an arbitrary scale. As a consequence, Waddington's landscape cannot be used to derive specific predictions about developmental mechanisms or their genetic control. Despite its geometric structure, the landscape model is non-mathematical, inexact, and qualitative. Its role is not to predict or explain, but to speculate and explicate. The restriction to three dimensions allows intuitions shaped by everyday experience to be brought to bear on aspects of development, genetics, and evolution as represented in the model. The price of this intuitive picture of canalization and other evolutionary developmental processes is rigor and precision. As few molecular details were available at the time, this was an easy trade-off for Waddington. The landscape model was intended for "a context in which it is more important to employ a system of thought which is flexible and of wide application than to search for a precise formulation of a narrower viewpoint" (1957, 31). To play its unificatory role, the model was constrained to be simple, rather than an accurate or principled representation of animal development.

To sum up: Waddington's landscape offers a convenient diagrammatic framework for conceptualizing the role of genes in development. It is simple, bears a clear relation to experiments, and represents genetics and development as complementary approaches, unified by structural correspondence. It also represents genes as controlling development: the fixed ground of developmental potential. As argued above, this last feature is no longer applicable. Modifications to the landscape introduced by cell reprogrammers bear out this claim.

6.6   Reprogramming and the landscape

Unusually for a product of the 1950s, Waddington's landscape appears in several high-profile reviews and commentaries on reprogramming.[133] Reprogrammers use the model to visualize shared background assumptions, express generalizations about experiments, and correlate cell state and developmental potential. These uses highlight several contrasts with Waddington's original model. First, stem cell biologists interpret the developing entity as a cell. Given this assumption, which Waddington allows for, but does not make himself, the landscape model visualizes key features of the stem cell concept: a single undifferentiated starting point, with the potential to develop along a variety of pathways, gradually restricted as development proceeds, and ending with stable, mature cell types.

A second, related contrast concerns 'units' of stability and determination. For Waddington, these are developmental pathways: "it is the track as a whole which, compared with any line lying between the tracks, is a description of an equilibrium" (1940, 92). He explicitly denies that points on the landscape represent equilibrium states of tissues. Instead, networks of interacting genes and gene products determine robust *pathways*, and pieces of tissue develop via a sequence of robust tracks punctuated by binary 'decisions.' Stem cell biologists, in contrast, take points on the landscape to represent cell states, which are more or less stable with respect to intervention. On this interpretation, the rolling ball represents a cell passing through different states in a process of differentiation. Under these assumptions, Waddington's argument that genes collectively determine the form of developmental pathways no longer applies. Moreover, the representation of gene action does not structurally correspond to that of development. These altered representational assumptions omit Waddington's rationale for the hypothesis that genes are the underlying determinants of development.

Another contrast concerns the relation of development and evolution. Stem cell biologists are, for the most part, unconcerned with evolutionary processes. Instead, they focus on specific mechanisms operating during an organism's lifetime or in the transparent, simplified 'bodies' of cell culture. They do not attempt to explain long-term changes in organismal populations, nor the gradual sculpting of adaptations, nor interspecies relations. So stem cell biologists do not share Waddington's rationale for treating the landscape as malleable. However, the landscape could be conceived dynamically on developmental timescales. Changes in topography could be induced by cell movement, experimental manipulation, or random 'noise' in cellular systems. Importantly, DNA sequences can today be altered as easily as other components of developmental mechanisms. So there is no reason to represent genes as uniquely stable or fixed. Instead, stem cell biologists treat the *entire landscape* as a fixed background for representing changes in cell state and potential (Figures 6.5 and 6.6).

In recent diagrams, the landscape provides a backdrop for generalizations about reprogramming experiments and hypotheses about their relation to normal development. Developmental processes are depicted as arrows describing trajectories on the landscape. In Figure 6.5, for example, normal development is visualized as a trajectory down the landscape, reprogramming as the reverse. More elaborate summaries compare and contrast different reprogramming experiments. In Figure 6.6, Pathway 1 shows "complete" reprogramming: conversion of a differentiated cell to a stable pluripotent state. Pathway 2 depicts "incomplete" reprogramming, in which a differentiated cell is temporarily converted to a pluripotent

(a) Development

(b) Pluripotent reprogramming
(SCNT, iPS)

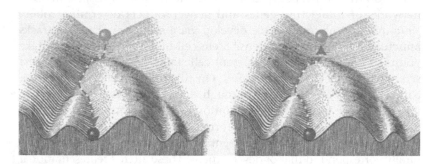

Figure 6.5 Waddington's landscape co-opted to represent reprogramming experiments. (a) Normal cell development as a trajectory down the landscape. (b) Reprogramming visualized as a reversal of normal development. Reprinted from Zhou and Melton (2008) with permission of Elsevier Press

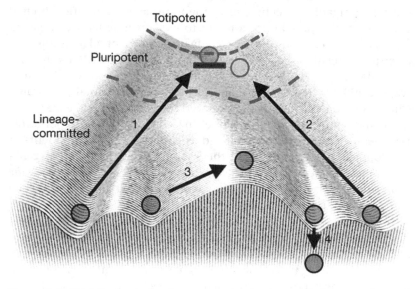

Figure 6.6 Waddington's landscape as the background for experimental manipulations of cell fate and potential. Reprinted from Yamanaka (2009) by permission of Macmillan Publishers Ltd: *Nature*

state, but normal development then resumes. Pathway 3 represents "transdifferentiation:" direct conversion from one differentiated cell type to another without passing through a pluripotent state. Finally, Pathway 4 depicts cell death – a frequent outcome of experimental manipulations, and also a 'programmed' response of cells in a variety of circumstances. Each pathway is a departure from normal development.

These modified landscape diagrams represent generalizations about reprogramming experiments. Just as Waddington generalized from experimental manipulations of embryonic development in fly, frog, and chick, abstracting details to highlight formal features of developmental processes, cell reprogrammers generalize from experimental manipulations of cell development. But while Waddington inferred formal features of organismal development from the results of experiments, stem cell biologists infer formal features of experimental methods themselves as manipulations of normal development. Different classes of reprogramming experiment are distinguished not by methodological details, but by the tracks they describe on the landscape; each a different deviation from normal development. In this way, the landscape model compresses thousands of reprogramming experiments into a few generalizations. Visualizing these on the landscape suggests possible explanations for experimental outcomes.

One generalization that has emerged from six years of reprogramming is that the success rate within an experiment (i.e. percentage of treated cells that become pluripotent) is inversely correlated with the source organism's age. For example, producing iPSC from cells of an adult is far less efficient than producing ESC from cells of an embryo. The modified landscape diagrams sketch an explanation: reprogramming "pushes" cells up the developmental incline, reversing the paths traced by their prior development (Figure 6.5). The source of the push is unspecified: it could be the experimenter, specific proteins, specific genes, or some combination thereof. The hypothesis represented is that the more differentiated the cell, the further "uphill" it must travel to reach a pluripotent state and the more likely it is that some other factor will block its path. Another experimental generalization is that pluripotency is unstable in most cells; most reprogramming is "incomplete." Reprogrammers speculate that complete, stable reprogramming requires an "epigenetic bump" to prevent the cell from rolling back down the hill once the inducing stimulus, whatever its nature, is removed (Yamanaka 2009, 50). The need for the extra "block" is shown as a dark bar on the landscape – a tiny 'black box' (Figure 6.6).

More detailed hypotheses elaborate on these sketches. But, strikingly, none appeal to the underside of Waddington's landscape, with genes as the "ultimate determinants" of developmental topography. Indeed, DNA sequences are seldom discussed. Reprogrammers look instead to other components of molecular mechanisms to explain generalizations represented on the landscape. One idea is that chemical modifications of DNA and chromosomal proteins progressively accumulate during development (see, for example, Zhou and Melton 2008, 386). The greater the portion of the epigenome needing to be "wiped clean," the greater the difficulty and the less likely the process is to succeed. Another proposal is that regulatory binding sites on nuclear DNA randomly shift from "open" to "closed" positions and vice versa. The few cells induced to pluripotency are just those with the right regulatory binding sites "open" at the right time (Hochedlinger and Plath 2008, Yamanaka 2009).

Intuitions underlying these proto-explanations are complicated, as background assumptions about development are entangled with twentieth-century gene-centrism. The basic principles of development, as visualized in the landscape, include unidirectionality and progressive restriction of developmental potential. These principles preserve the traditional idea that development is irreversible, presupposed in much embryological thought, and carried forward into developmental biology. The notion that genes are fundamental for biological explanation encourages the view that genetic changes to nuclear DNA are irreversible (Keller 2002). All other kinds of changes, lumped together as "epigenetic," must then be reversible, at least in principle. The process of organismal development (for the most part) consists of epigenetic changes, as early reprogramming experiments demonstrate. Reprogramming is thus the artificial exception that proves the normal rule, that development is irreversible, by exploiting its in principle reversibility. Entrenched association of epigenetics with reversibility of development makes it intuitively plausible to suppose that reprogramming is just development in reverse, as depicted on the landscape. But this assumption rests on ideas about genetic primacy that are not generally accepted in stem cell biology. So particular care is needed when interpreting reprogramming experiments, to avoid bias in favor of this unmotivated supposition. With this caveat in mind, Waddington's landscape sets the stage for MEx of developmental phenomena more generally.

This is accomplished in the model's third use by stem cell biologists: correlating cell state and developmental potential. As discussed in previous chapters, stem cell experiments measure both cells' molecular traits and their developmental potential. An array of such experiments correlates the two sets of measurements. Waddington's landscape is a helpful device for visualizing this correlation (Figure 6.7). The higher

Developmental potential	Epigentic status

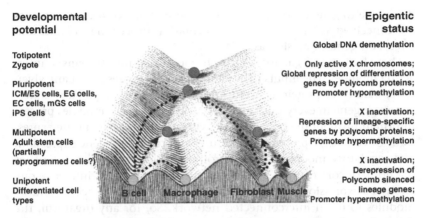

Developmental potential

Totipotent
Zygote

Pluripotent
ICM/ES cells, EG cells,
EC cells, mGS cells
iPS cells

Multipotent
Adult stem cells
(partially
reprogrammed cells?)

Unipotent
Differentiated cell
types

Epigentic status

Global DNA demethylation

Only active X chromosomes;
Global repression of differentiation
genes by Polycomb proteins;
Promoter hypomethylation

X inactivation;
Repression of lineage-specific
genes by polycomb proteins;
Promoter hypermethylation

X inactivation;
Derepression of
Polycomb silenced
lineage genes;
Promoter hypermethylation

B cell   Macrophage   Fibroblast   Muscle

Figure 6.7 Correlation of cell state and developmental potential via the landscape model. Reprinted from Hochedlinger and Plath (2009) with permission from *Development*

a cell's position on the slope, the greater its developmental potential. This position, representing cell state, can then be associated with a pattern of gene expression, constituted by interacting DNA, RNA, protein, and small molecules. Experimental manipulation of the molecular components of regulatory networks reveals the details of these mechanisms. The landscape model thus offers a convenient coordinating framework, to be filled out by further experiments.

6.7 Regulatory genes

So far, I have argued that, in experimental biology in general and stem cell biology in particular, genes are better understood as vital components of complex molecular mechanisms than "master molecules" controlling development. In this final section I consider an influential opposing view: "the regulatory genome" (Davidson 2006, Davidson and Levine 2008). On this view, development is controlled by a DNA-encoded program made up of short sequences that specifically bind TF proteins and thereby make a difference to gene expression. Davidson terms these DNA sequences "cis-regulatory modules." Each module's effect is represented as a conditional rule of the form: "If protein X is present, then gene Y is expressed at level Z."[134] Each gene has a set of such modules associated with it, which collectively specify its expression pattern under various conditions. So control of development is attributed entirely to the DNA components of molecular complexes that make a difference to gene expression. Protein components of these complexes are conceived as "inputs" to the information-processing modules and effects on gene expression as "outputs."

These basic "cis-regulatory" units are organized into systems of interacting modules. The inputs that determine gene expression at any given time, mainly TF protein concentrations, comprise a "regulatory state." For every gene in every nucleated cell, regulatory DNA modules process information about the cell's regulatory state into effects on gene expression. Of course, TF proteins are themselves products of gene expression. The regulatory modules that control expression of TF genes are sites of "primary core control" for development. Because TFs influence one another's expression, both positively and negatively, the "core control" modules form an interconnected network. So, for any organism, the genetic regulatory network (GRN), composed of DNA sequences distributed throughout its genome, constitutes a stable underlying program for development.

DNA, though not genes in the traditional protein-encoding sense, therefore explains development:

> The design features of the GRN directly explain why the events of a given process of development occur; for example, why a given set of cells becomes specified to a specific fate, why it emits particular signals to adjacent cells, and why it differentiates in a given direction. The architecture of a GRN is mandated by the *cis*-regulatory sequences...that control each gene of the network. These sequences determine what inputs affect expression of each gene, and how these inputs operate in a combinatorial fashion.
>
> (Davidson and Levine 2008, 20063)

Collectively, DNA modules act as "a vast, delocalized computational device" that processes regulatory states of a cell (Davidson 2006, 185). These sequences are (for the most part) invariant across cells of an organism and organisms of a species, and a relatively small set of TF is conserved throughout much of the animal kingdom. GRNs, therefore, offer a unified explanation of animal evolution and development.

The regulatory genome is an explanatory model based on experimental results accumulated over several decades, which revealed core mechanisms of early development in a few model organisms (notably sea urchin and *Drosophila*). Essentially, Davidson and colleagues interpreted this enormously detailed and complex dataset in light of the assumption that DNA controls development. By omitting or drastically simplifying other components, they produced a more tractable model, which they then extrapolated to animal development in general. There is nothing problematic about this modeling approach. The question here is whether the basic assumption, that DNA controls development, is justified in stem cell biology. Several points made above support a negative answer. First, gene expression, which is central to explanations of stem cell phenomena, involves many components besides DNA. Because DNA sequences are (for the most part) invariant across cells of an organism, they cannot be actual difference-makers for gene expression. So there is no reason to privilege DNA as a cause of development. GRN models actually reflect this: features of non-DNA components that make a difference to gene expression are represented in compressed form, as arrows linking regulatory DNA sequences. The causally active components of GRNs are not limited to DNA sequences.

Nor can DNA claim causal priority in virtue of being the initial step in the linear chain, DNA→RNA→protein. As a causal model, the Central Dogma has been superseded by interactive networks in which diverse TF are implicated in the expression of any particular gene, TF proteins have context-dependent effects, and micro-RNAs influence gene expression and protein activity. Within such a network, no single position can be non-arbitrarily identified as the start-point. It follows that DNA cannot occupy such a position. Even at the earliest stages of an organism's existence, DNA, RNA, proteins, and small molecules are organized in elaborate, ongoing interactions. Nor is DNA set apart from other network components by its informational properties. GRNs do not involve coding relations like those linking terms of the Central Dogma, namely that DNA sequence is a template for RNA sequence, which is, in turn, a template for an amino acid sequence. The linguistic terms 'transcription' and 'translation' refer to the molecule-by-molecule mapping that is conceptualized as a code. In contrast, regulatory modules in genomic DNA specifically bind proteins not as templates that allow mapping between sequences, but via molecular bonds mediated by conformational and electrochemical properties – just as complexes of proteins and RNA do.

So the role of DNA sequences in gene expression is not distinctive, either causally or informationally. However, DNA is distinguished from other components of developmental regulatory mechanisms by invariance within an organism. DNA sequences can therefore provide a stable, underlying architecture for development. For Davidson, committed to a unified explanation of evolution and development, this is a crucial point. Regulatory DNA sequences differ across species and so are actual difference-makers for development in comparative perspective. These explanatory aims motivate the assumption of genetic control. But stem cell biology is concerned with the lives of individual organisms within species. The two explanatory projects are distinct. Davidson's account is intended to provide a unified explanation of development and evolution, with particular emphasis on variation among species, and focuses on development of the whole organism. Stem cell biologists, in contrast, aim to explain development at the cellular level, with particular emphasis on variation within an organism. The regulatory gene model therefore plays a minimal role in explanations of stem cell phenomena, while stem cell phenomena are explicitly omitted from Davidson's account.[135]

The aims of stem cell biology do not rule out attributing explanatory importance to invariant DNA sequences within an organism. Their explanatory role cannot, however, be one of causal control. Instead, stem cell biologists might privilege regulatory genes as stable *coordinators* of diverse molecular interactions. The controlling gene would then give way to the coordinating gene in explanations of stem cell phenomena.

6.8   Conclusion

This chapter criticizes the view that genes have a privileged role in explanations of development. The causal-informational gene concept has been criticized before, in light of the paradox of development and the rising tide of molecular complexity. However, focus on stem cells brings new dimensions to the debate. Philosophical accounts of genes in development tend to frame the issue in terms of molecules and organisms, neglecting the mediating levels of cells, tissues, and organs. The latter, however, comprise the major domains of stem cell phenomena. In particular, iPSC and cell reprogramming offer a valuable test case for Waters' account of genes as privileged difference-makers in development. Close examination shows that DNA sequences are *not* actual specific difference-makers in stem cell experiments. Instead, reprogramming experiments aim to reveal interacting, joint causes of developmental pathways. These results further support the joint account of MEx, presented in Chapter 5, which is incompatible with genetic explanatory privilege.

These negative arguments motivate an alternative account of the role of genes in development, which is provided by Waddington's landscape. The latter model is particularly apropos, as it was originally intended to integrate ideas about genetics and development and has recently been co-opted by stem cell biologists to represent experimental results and speculative explanations thereof. Waddington's original model had three purposes: to represent developmental pathways, express the hypothesis that genes indirectly control development through a network of interacting biochemical products, and 'visually unify' genetics, development, and evolution. Stem cell biologists update the landscape model to serve somewhat different purposes, one consequence of which is "parity" of genes with other components of gene expression mechanisms. Waddington's model, with these modifications, offers a framework for MEx of cell development. Finally, an influential alternative, which identifies a subset of DNA sequences as the "core program" controlling development, is shown to be peripheral to stem cell biology today.

7
Pluripotent Model Organisms

7.1 Introduction

Embryonic stem cells were not the first cultured stem cells. That distinction belongs to 'embryonal carcinoma cells,' first derived in the 1960s from mouse cancer cells. Approximately a decade later, embryonic stem cells were derived from mice. But it was nearly 20 years before a similar method was successfully applied to humans. *Human* embryonic stem cells (hESC) are a recent innovation. Today they symbolize the hopes of regenerative medicine, spurring reorganization of biomedical resources and reformulation of research goals to encourage new forms of collaboration, new technologies, and new ways of engaging in politics.[136] In this chapter, I argue that their role is best characterized as that of a model organism, selected and modified to serve as an exemplar for biological research. I show how hESC were originally constructed out of earlier model systems, tracing this line of research back to studies of cancer in the 1950s. Over five decades, a network of models was constructed via complex comparative relations among pluripotent cell lines. The origins and structure of this 'pluripotency network' shed light on both the organization of stem cell research today and the significance of hESC in these biomedical inquiries.

7.2 Model organisms in experimental biology

Model organisms play a prominent role in contemporary life science. Since the early twentieth century, biologists have concentrated their research efforts on a few select organisms: the fruitfly *Drosophila melanogaster*, the nematode worm *Caenorhabditis elegans*, the laboratory mouse *Mus musculus*, *Escherichia coli* bacteria, and *Saccharomyces cerevisiae* yeast, among others. These models are designed to represent general biological phenomena, and their traits reflect our purposes, as well as their own material nature. They also play a dual epistemic role, being objects of study in their own right on the one hand, and tools for studying other biological systems ("targets") on the other. This dual role is characteristic of scientific models in general (Morgan 2007, 264–270). However, models in science are so diverse that little else can be said of them generally. Philosophical understanding of models and modeling comes from detailed consideration of cases.[137] Case studies of model organisms have yielded many insights about experimental biology and its relation to wider social contexts. From these studies, four robust points have emerged.

First, model organisms are initially selected as research objects on the basis of practical considerations: they are cheap, fast, easy to work with and can be raised in large numbers. Organisms that meet these practical criteria tend to share certain features: small body size, large offspring number, rapid development to sexual maturity, and robust processes of reproduction and development. These features are nearly universal among model organisms in biology, but not organisms in general. This is but one instance of the modeling role introducing subtle, but potentially problematic, biases in the model–target relation (see §7.3). Second, model organisms that ground successful research programs exhibit biological phenomena of interest in a way accessible to study. Often this means that a particular biological feature is exaggerated in size, such as the squid giant axon and *Drosophila* chromosomal "puffs." Increased visibility is also prized; organisms with transparent bodies offer 'windows' into physiology and development. Above all, model organisms should be simple enough that phenomena of interest can be manipulated by controlled experiments that yield robust and clearly interpretable results. In developmental biology, this requirement led to choice of organisms with an early separation of germline/soma and highly canalized developmental processes robust to environmental change (Bolker 1995, 451–452, Gilbert 2001, 3). For both practical and epistemic reasons, model organisms are a decidedly non-representative sample of organisms.

Third, once introduced to laboratory life, these select organisms are altered to better suit researchers' purposes. Robust experimental results are rarely obtainable from organisms in natural populations, particularly for phenomena as complex as development. So modifications are introduced in the laboratory to render organisms more tractable. In this sense, successful model organisms are 'constructed.' Construction of model organisms typically involves a breeding program and a constant, highly artificial environment, both of which reduce variation among individual organisms within a laboratory stock. Other modifications are introduced to bring reproducible results in line with general explanations or theories. In one important example, Morgan's Fly Room at Columbia, the two kinds of modification coincided: strains of *D. melanogaster* were engineered so as to exhibit principles of classical genetics (Kohler 1994, Waters 2007, Weber 2007). Manipulations of breeding stock decreased background variation *and* showcased mutations conforming to Mendelian inheritance. Fourth, laboratory modifications are not unlimited, but constrained by the material nature of the organisms themselves. Model organisms' "recalcitrance" is as scientifically important as their tractability. Adaptations to a laboratory environment go beyond experimenters' intentions. These novel traits can, in turn, spur new experimental efforts. Other features may not be amenable to change at all. So results of model organism research emerge from the interplay of biological nature and engineered manipulation. Their insistent material life has profoundly influenced the course of twentieth century biology, while adaptation to laboratory environments and researchers' purposes has, in turn, greatly altered these organisms.

7.3   An epistemic challenge

The above processes systematically differentiate model organisms from their naturally-occurring counterparts, producing an epistemic trade-off between the two aspects of their epistemic role. Owing to their simplicity, tractability, and so on, knowledge about model organisms accumulates rapidly. But these same features make model organisms dissimilar to their apparent representational targets. Model organisms are exceptional, with traits reflecting both researchers' aims and pressures of the laboratory environment. As noted above, the very features that render them useful for research introduce systematic biases that undermine inference to conclusions about non-model organisms. This is a problem whether conclusions are general or particular. A small sample of atypical species is inadequate ground for generalization about large groups of taxa.[138] So generalizations from model organism development, say, to development of all animals (or large groups, such as bilaterians or vertebrates) should be treated as highly speculative. Similarly, extrapolation to particular non-model organisms (notably humans) is often suspect. Such extrapolation can be underwritten by strong causal analogies between model and non-model organisms. But usually too little is known about non-model organisms to support such analogies; if this were not so, the model organisms would be superfluous. The epistemic asymmetry is such that model organisms cannot directly license inferences about non-model organisms. Instead, their role is to suggest hypotheses, which must be further tested.

These considerations help clarify model organisms' epistemic role. Experiments on model organisms, if properly performed, yield knowledge about biological phenomena *in those particular organisms*. But these results do not license direct generalization or extrapolation to claims about non-model organisms. Though the latter may be targets of *inquiry*, they are not model organisms' primary representational targets. Instead, model organisms represent biological phenomena of interest in simplified, tractable form – just as abstract models do.[139] Model organisms differ from their abstract counterparts in having a material aspect that eludes complete understanding and control. But their modeling role is the same: to provide a platform for comparison, furnishing potential points of analogy and disanalogy with which to conceptualize targets of inquiry. For this role, the systematic biases in model organism construction are not a problem, but a resource. In particular, model organisms' distinctive features provide a starting point for comparative analysis. Case studies of *C. elegans*, tobacco mosaic virus, and the Human Genome Project support this interpretation of model organisms' role (Ankeny 1997, 2001, Creager 2002). For example, tobacco mosaic virus suggested points of analogy for study of other viruses and subcellular structures, rather than being taken as a direct representation of the latter. Important general properties of viruses and microsomal particles were worked out in relation to one another. Stem cell research, I will argue, exhibits a similar pattern.

Before turning to stem cells, however, a few key terms need clarification, notably "model system" and "model organism." Though biologists often use the two interchangeably, it will be helpful to distinguish them here, as well as the related notion of "experimental system." An experimental system consists of instruments, techniques, materials, and skilled practitioners organized into a knowledge-producing assemblage that is not guided by theory (Rheinberger 1997). The material aspects of such assemblies have an inherent capacity to surprise us, direct research along unanticipated paths, and spur reformulation of aims and core concepts. Rheinberger's account of experimental systems focuses on stages of research in which conceptions of the objects under study ("epistemic things") are in flux. Model systems come into play when concepts are more settled, and an experimental system is taken to yield reliable knowledge about the objects under study, albeit in a restricted context. But the open-endedness Rheinberger emphasizes is retained in the modeling role, when reliable knowledge about objects in the model system is speculatively extended to new targets of inquiry. So model systems in use are a kind of experimental system, going beyond settled boundaries to yield knowledge of other objects – new epistemic things. Experimentation on model systems thus involves continuous reshaping of materials, methods, aims, and concepts.[140] Similarities and differences across experimental systems are clarified as part of this ongoing process. A successful model system is a productive exemplar for other experimental systems.

Model systems in biology consist of organisms or their parts together with experimental techniques and methodologies used to manipulate them. So model organisms are key parts of (some) model systems. But, as concrete entities in their own right, model organisms can also separate from a surrounding model system and circulate as "boundary objects," facilitating collaboration among different communities. For example, exchange of genetically standardized organisms, notably fruitflies and mice, contributes to consensus formation in many biological fields (Kohler 1994, Rader 2004). Traffic in model organisms can also reveal important social dimensions of experimental biology. For example, wide use of tobacco mosaic virus as a model in the mid-twentieth century was achieved, in part, by political initiatives and changes to biomedical infrastructure. Once in common use, this model forged links between biochemistry and molecular biology laboratories, generating new technologies and concepts (Creager 2002, 317–333). Recently, social scientists have studied human embryos as boundary objects, facilitating linkage of the social worlds of embryonic stem cell research and pre-implantation genetic diagnosis (Williams et al. 2008).[141] "Mapping" projects of this kind might fruitfully synergize with case studies of model organisms in genetics, developmental biology, and cancer research.

The following sections take a more modest approach, focusing on model organisms as parts of model (and experimental) systems. As the examples mentioned illustrate, studies of model organisms link the material culture of experimental practice with wider social and cultural contexts. However, scholars of stem cell research focus overwhelmingly on the latter, with particular attention to political and ethical controversies over use of human embryos in research. Comparatively little attention has been paid to stem cell experiments in their narrower scientific context. The model organism perspective can also illuminate this. Stem cell biology uses many classic model organisms, such as flies, mice, and bacteria. But a distinctive component of its experimental systems are cultured stem cell lines. These, the next section argues, also qualify as model organisms.

7.4 Stem cell lines as model organisms

Cultured stem cell lines are laboratory artifacts. The basic method for producing them is described in Part I. To briefly recap: cells are removed from an organismal source and placed in artificial culture, with chemical factors added to prevent differentiation. Often, extracted cells are spread on a layer of "feeder cells" that supply additional components of a stem cell microenvironment, or "niche." Cells that rapidly divide under these conditions form colonies, which are selected and put into new cultures. Repeated "passaging" every few weeks maintains a continuously-growing lineage of undifferentiated cells. Such a cell line is constructed to be self-renewing, so only differentiation potential needs to be experimentally demonstrated to establish that it is a stem cell line (with the caveats noted in Chapter 3). Continuously growing lines of undifferentiated cells which can differentiate under appropriate culture conditions qualify as stem cell lines. The many types of stem cell cluster into two main branches: tissue-specific and pluripotent. Tissue-specific stem cells are extracted from fully- or partly-developed organisms, are difficult to culture outside the body, and exhibit limited self-renewal and differentiation potential. Pluripotent stem cells, in contrast, are extracted from organisms in very early stages of development, are maintained as cell lines in culture, and exhibit unlimited self-renewal and very broad differentiation potential. Most cultured stem cell lines are, therefore, pluripotent.[142]

Cultured stem cell lines exhibit all the characteristic features of model organisms: small body size, large offspring number, rapid growth rate, robust processes of reproduction and development, tractability, simplicity, and accessibility to observation and measurement. The body size of a cell line is flexible in principle, but, in practice, constrained by the dimensions of dishes and flasks – typically about the size of one's hand. Multiple cell lines can be maintained in a single incubator the size of a small refrigerator. Continuous growth is obviously a hallmark of pluripotent stem cell lines. Cells in these cultures are selected for frequent division; passaging about every fortnight ensures a rapid reproductive rate. Numbers of individual cells increase geometrically nearly every day unless deliberately culled by experimenters. So cultured stem cell lines exhibit both large offspring number and rapid growth. They also give biologists unprecedented access to phenomena of mammalian development. Cultures grow in clear plastic containers: "transparent bodies" that make all cells in a culture visible and exposed to external manipulation.[143] This unlimited access contrasts strikingly with normal mammalian development, which is sequestered within an adult's body and all but undetectable in early stages.

Stem cell lines do vary in their tractability; mouse stem cells, for example, are much easier to work with than their human counterparts. As new stem cell lines are created and new methods pioneered, researchers continue to streamline and standardize stem cell culture conditions. Once a new stem cell line is established, it is modified to reduce variation among constituent cells and further manipulated so as to yield reproducible experimental results. As experimental identification of stem cells depends crucially on homogeneous cell populations (see Chapter 3), reducing variability among cells within a line is a high priority for any laboratory. Larger-scale standardization efforts are designed to increase efficiency in collaborations and accelerate progress toward clinical results (e.g. Adewumi et al. 2007, Bock et al. 2011, Loring and Rao 2006). These efforts resemble standardization practices for classic model organisms, such as inbred mice and fly strains. So, like classic model organisms, cultured stem cell lines are constructed to be tractable for laboratory use and to exhibit phenomena of interest (differentiation and self-renewal) in a way accessible to controlled manipulation. Finally, as objects of experiment, stem cell lines are concrete objects of investigation in their own right. They are also tools for representing early development and cell differentiation in mammalian organs and tissues.[144] So stem cell lines fulfill the dual epistemic role of a model organism.

However, stem cell lines do differ in some striking ways from the classic model organisms of developmental biology: *D. melanogaster, C. elegans*, mouse, chick, the frog *Xenopus laevis*, and zebrafish (Gilbert 2001). The latter all share some significant developmental traits: rapid development to sexual maturity, early separation of germline from soma, and highly canalized developmental processes robust to environmental change (Bolker 1995). Stem cell lines exhibit none of these. Cell division is continuous and involves no sexual reproduction. Accordingly, stem cell lines show no development to sexual maturity and no germ/soma separation. And far from being canalized, stem cells are exquisitely sensitive to their environments. These contrasts do not disqualify stem cell lines as model organisms for developmental biology, but are instead clues to their distinctive representational role. Stem cell lines are extremely simple, lacking any sexual aspect or germ/soma distinction. In place of an early embryo's multilayered three-dimensional structure and complex environment are a transparent artificial 'body' and simple fluid medium of cell culture. All these features hint at a key modeling role: stem cell lines embody early mammalian development in drastically simplified form, reducing the complexity of an organismal body to a layer of undifferentiated, dividing cells selected for homogeneity. Their context-sensitivity is the cornerstone of differentiation experiments, which manipulate culture environments and compare cell traits before and after the change. Putting the two features together, stem cell lines exhibit cell development in a very simple and manipulable way – exactly as a model organism should.

As objects of experimentation and embodiments of simplified mammalian development, stem cell lines occupy the dual epistemic role of model organisms. I conclude that cultured stem cell lines *are* model organisms. But this claim requires some defense. One objection is that cultured cell lines are not organisms, but explanted *parts* of organisms. As both can be parts of model systems, the distinction is not crucial here.[145] But on a common-sense understanding of organisms, the objection does not hold up. Granted, stem cell lines do not greatly resemble the organisms from which they derive. This is a consequence of their extreme simplicity compared with multicellular animals: precisely what makes them useful models for studying mechanisms of mammalian development. But stem cell lines are not so simple that they cannot be considered organisms at all. Microorganisms, after all, are organisms too. Classic model organisms include strains of bacteria and yeast, which exist as single cells. Stem cell lines are derived from multicellular organisms but resemble microorganisms in their organization and reproduction.[146] Like more familiar organisms, they grow, reproduce, evolve by selection and drift, and occasionally perish from disease or inhospitable environments. Unlike classic model organisms, stem cell lines are artificial; their activities depend on human technology and intervention. But this means only that they are not organisms *found in nature*, not that they are not organisms. Many model organisms cannot survive outside a laboratory environment. For example, certain mouse strains, widely used in biomedicine, lack an immune system, and must be maintained in a sterile room. Being a highly-engineered obligate inhabitant of a laboratory does not disqualify an entity from being an organism. The "synthetic cell" produced by Venter and colleagues is often characterized as an organism with a "minimal genome," capable of reproduction and growth in a laboratory environment (Gibson et al. 2010). Stem cell lines are, analogously, multicellular organisms with 'minimal development,' embodying essential developmental processes in artificially-simplified form.

Another objection arises from the idea that organisms are defined by species membership. Stem cell lines belong the same species as their source (*M. musculis* or *Homo sapiens sapiens*). But they are manifestly not organisms of those species, lacking the distinguishing characteristics of mammals and vertebrates, and bilateral or even multicellular organization. It is tempting to conclude from this that stem cell lines are not organisms and therefore not model organisms. The objection fails, however, because the starting premise does not hold up in experimental biology. The species category is too coarse-grained to capture distinctions among organisms that are significant for biological practice. Consider the diversity of model organisms within *M. musculis*: inbred strains, human–mouse chimeras, "knockouts" for specific genes, and so on. To understand the roles of these constructs in experimental biology, analogies and contrasts within, as well as among, species must be taken into account. Many case studies of model organisms do not emphasize this point, because they focus on the initial transfer from natural population to laboratory.[147] At this stage, species classifications are often important: one species is chosen over others to be a model organism. But stem cell lines were created after classic model organisms were already established. At this stage, proliferation of models within and among laboratories does not hinge on species distinctions, but finer-grained comparisons and contrasts. The next section examines these relations in detail. As for other model organisms, to understand the epistemic role of stem cell lines we need to examine how such models were originally constructed and modified. Here, history of stem cell research comes to the fore.

7.5   Origins of pluripotency research

The history of stem cell lines falls into three stages: teratocarcinoma research (1960s–1980), genetic studies (1981–1998), and contemporary pluripotency research (1998–present). Each exhibits a distinctive array of modeling relations, which frames the next stage. Transitions are marked by creation of key exemplars: embryonic stem cell lines from mice (1981) and humans (1998). The sequence culminates in the complex representational network of pluripotency research today. The timeline of this branch of stem cell biology is summarized in Table 7.1.

7.5.1   Embryonal carcinoma

Pluripotency research began when phenomena of self-renewal and differentiation became tractable for study through research on a classic

Table 7.1 Pluripotency research timeline

1954	Teratocarcinoma-129
1964	Embryoid bodies
1970–74	Mouse EC (many lines)
1975–77	Human EC (few lines)
1981	Mouse ESC
1987–89	Knockout mice
1992	Mouse GSC
1995	Monkey ESC
1998	Human ESC, human GSC
2004	~~Cloned hESC~~ (fraudulent)
2006	Mouse iPSC
2007	Human iPSC, mouse epi-SC

model organism: inbred mice. One strain ("129") showed an unprecedentedly high incidence of teratocarcinoma in testes (~1% of males; Stevens and Little 1954). Teratocarcinoma is a cancer that manifests as tumors containing diverse cell types: fat, muscle, skin, nerves – even teeth. These tumors occur spontaneously, though rarely, in testes of humans and horses, and can also be experimentally induced. Strain-129 mice, unexpectedly, were a reliable source of teratocarcinoma, providing sufficient material for study under controlled conditions. These experiments showed that strain-129 teratocarcinomas contained undifferentiated cells, which gave rise to more specialized cells. Strain-129 teratocarcinoma was the first pluripotent stem cell line.

Construction of this cell line was guided by an analogy between teratocarcinoma and embryonic development. The former was conceived as a simplified version of the latter, in which diverse "normal-type tissue cells... stem from embryonic undifferentiated cells" (Stevens and Little 1954, 1086). Undifferentiated cells within a tumor were variously described as "embryonic," "embryonal," or "embryonic-type." Like early embryos, undifferentiated teratocarcinoma cells give rise to a wide range of cell types, including products of "each of the three embryonic germinal layers" – the operational criterion of pluripotency (Kleinsmith and Pierce 1964, 1544). However, the organization of embryonic development was lacking: teratocarcinomas "appear to recapitulate many of the events that occur during early embryonic development but in a disorganized manner," yielding "a bizarre neoplasm composed of foci of undifferentiated malignant cells interspersed with a chaotic array of somatic tissues" (Andrews 2002, 406, Kleinsmith and Pierce 1964, 1544). Another dissimilarity is that teratocarcinoma "embryonic-type cells" continually self-renew, persisting alongside more differentiated tumor cells. These "prolific undifferentiated cells" are the source of malignancy.[148]

The analogy between teratocarcinoma and embryonic development was elaborated further by two kinds of experiment: transplanting cells into mice, and growing them in culture. This dual experimental approach persists in stem cell biology today. Transplanted teratocarcinomas produced more tumors in new hosts. One "rapidly-growing transplantable tumor" consisting of "prolific undifferentiated cells" was propagated through many generations of mice and split into several distinct lines (Stevens and Little 1954, 1084). This tumor cell line (402A) became an independent biological entity, with 129-mice its controlled environment. Another new experimental entity extended the embryo–cancer analogy even further. When suspended in fluid, undifferentiated teratocarcinoma cells aggregate into spheres ~0.1–0.2 mm in diameter with an inner core of undifferentiated cells wrapped in a layer of more differentiated cells resembling embryonic endoderm. These structures resemble embryos at the blastocyst stage, which are organized into three germ layers (ectoderm, endoderm, and mesoderm) with characteristic developmental fates. Experiments suggested similar patterns of developmental potential in the different layers of teratocarcinoma-derived "embryoid bodies."

To summarize: in this first stage, the analogy of cancer and embryonic development was explored by constructing a network of interrelated models (Figure 7.1). Though the experiments were in mice, observations of human teratocarcinoma were interpreted in the same framework.[149] Over the next two decades, the teratocarcinoma modeling network was used to reveal mechanisms of normal embryonic development. Again, research proceeded on two fronts: artificial cell culture and transplantation into mice. The former produced the first cultured stem cell lines; the latter, means of demonstrating pluripotency by experiment. Together, these teratocarcinoma studies established the basic design of stem cell experiments today.

The first cultured stem cell lines began with undifferentiated teratocarcinoma cells placed in artificial culture, which rapidly divided in culture and produced teratocarcinomas when transplanted back into strain-129 mice – showing both self-renewal and differentiation. In the 1970s, many of these "embryonal carcinoma" (EC) lines were produced to investigate molecular mechanisms underlying pluripotency and differentiation. Similarities accumulated between EC derived from mice (mEC) and early mouse embryo cells: morphology, surface molecules, and enzyme activity. These similarities encouraged the idea that mouse EC are a "malignant surrogate for the normal stem cells of the early

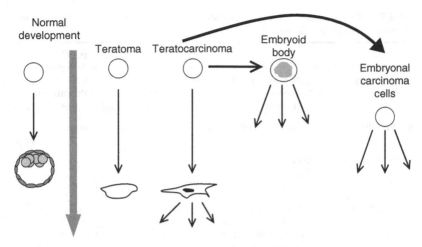

Figure 7.1   Teratocarcinoma models

embryo" (Andrews 2002, 405). The analogy was then extended to transplantation experiments in mice, with startling results: mouse *embryos* transplanted into 129-mice produced teratocarcinomas, just like mEC. The reverse experiment was then performed: transplant mEC into mouse blastocysts, then transfer the latter to uteri of 129-mice. The resulting offspring were chimeras: mixtures of cells from the original blastocyst and mEC. These experiments showed that mEC and early embryo cells not only share morphological and molecular traits, but can also play the same developmental roles. This functional equivalence validated mEC as "a model system for studying certain aspects of cellular differentiation" (Jacob 1978, 249).

But EC also had limitations as a model of development. Most problematic was their variable differentiation capacity. Some EC lines differentiated spontaneously, while others required a special environment. In vivo experiments showed even greater heterogeneity, as cells from the same embryoid body gave rise to different tissues in animals. Human EC did not differentiate at all under culture conditions used at the time; their developmental capacities went unrecognized for a decade. Comparison of cell traits in human and mouse EC was also complex and, at the time, lacked mechanistic explanation (Table 7.2). So caution was warranted:

The results obtained with teratocarcinoma should obviously be taken with some caution before being extrapolated to the embryo.

Table 7.2　Comparison of human and mouse EC (circa 1970s). ICM cells of mouse blastocyst resemble mEC for all traits listed

		hEC	mEC
Morphology	Cytoplasm/nucleus ratio	Low	Low
	Nucleoli	Prominent	Prominent
	Cell arrangement	Tightly-packed	Tightly-packed
Biochemistry	Alkaline phosphatase activity	High	High
Surface molecules	SSEA1	–	+
	SSEA3	+	–
	SSEA4	+	–
Development	Pluripotency	No	Yes
	– teratocarcinoma	–	+
	– embryoid bodies	–	+
	– chimeric mice	–	+
	– altered germline	–	~

Actually the two systems have to be used in a complementary way, each one being more adapted to certain experimental approaches. It is perhaps from a permanent interplay between the two systems that detailed analysis of early embryonic development may be expected to proceed.

(Jacob 1978, 266)

Pluripotency research today evinces just such "permanent interplay" among experimental systems, though teratocarcinoma and EC are not at the forefront. Their stage of pluripotency research came to an end with innovations involving the first embryonic stem cell line: mouse ESC.

Even these developments however, were foreshadowed in teratocarcinoma research. As noted above, as well as artificial culture, undifferentiated EC cells were propagated in tumors and embryoid bodies contained in mice. Their differentiation potential was assessed by transplanting "core" cells of an embryoid body into mouse blastocysts then observing tissue make-up of resulting offspring. Results were highly variable, as noted. However, one group, led by Beatrice Mintz, produced chimeric mice in which EC cells contributed to a wide range of tissues: blood, kidney, liver, hair, and – most strikingly – germ cells. This was "proof of principle" that mEC are pluripotent, though the conclusion could not be generalized to other EC lines without testing them directly. But Mintz also ventured another claim: pluripotent stem cells, "in conjunction with experimental mutagenesis" could be "a new and useful tool for biochemical, developmental, and genetic analyses of mammalian differentiation" (Mintz and Illmensee 1975, 3585). This idea was soon validated in spectacular fashion.

7.5.2 Mouse ESC

The second stage of pluripotency research began with a modest extension of teratocarcinoma research. As mEC shared many traits with cells in early mouse embryos, and experiments suggested that the two could play the same developmental role within mice, the next step was to culture an embryonic analog to mEC and examine its developmental capacities. This experiment added to "the network of inter-relationships between the mouse embryo and pluripotential cells derived from it [which] has previously lacked only the direct link between the embryo and cells in culture for completion" (Evans and Kaufman 1981, 155). The new mouse embryo-derived cell lines were designed with mEC as an exemplar. Strain-129 was the organismal source; the precise timing and location of cell extraction from mouse-129 embryos was based on molecular similarity to mEC; culture conditions were nearly identical; and growing cell lines selected for morphological and molecular resemblance to mEC (Evans and Kaufman 1981, Martin 1981; see Table 7.2). Resulting lines were then assessed by the same experiments that revealed mEC developmental capacities: production of teratocarcinoma in adult 129-mice, formation of embryoid bodies in liquid culture, and development of chimeric mice from blastocyst transplantation. The first cultured stem cell line was thus the model for the next (Figure 7.2).

At first, mESC were simply a new model of the teratocarcinoma research program, deepening the analogy between embryonic and tumor development by demonstrating that cells from an embryo can be manipulated to resemble undifferentiated teratocarcinoma stem cells. Though these connections made mESC a promising model organism for studying onset of cancer, a disanalogy with mEC instead dominated researchers' attention. As noted above, mEC differentiation is highly variable, both within and across cell lines. In contrast, mESC are reliably pluripotent, differentiating into a wide array of tissues as embryoid bodies, teratocarcinomas, or chimeric mice. When transplanted into 129-strain blastocysts, which were then brought to term in mouse uteri, mESC contributed to many tissues of resulting offspring – including the germline. Though Mintz and colleagues had produced similar results with mEC, their accomplishment was exceptional. With mESC, germline alterations in mice could be reliably produced. These results showed that developmental capacities

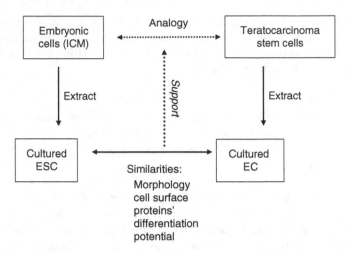

Figure 7.2   The main modeling relations of pluripotency research (1980s). Top: biological entities. Bottom: constructed models

of mEC and mESC are subtly different; though both are pluripotent in principle, the capacity is more robust in mESC.

It does not follow, however, that mESC is a *better model* for cell development than mEC. Rather, both belong to the same network of models used to probe early mammalian development via the teratocarcinoma analogy (Figure 7.2). Their relation is synergistic; the *network* of models is the better for including them both. But as a tool for genetic manipulation, mESC was clearly superior. As genes took center stage in developmental biology, this technical role eclipsed mESC's significance as a model. Teratocarcinoma models receded in importance, while mESC came to prominence in connection with a model system for investigating not cell development, but gene function in whole organisms – the knockout mouse.

The method for producing knockout mice combines mESC technology with gene targeting, an application of molecular biology. Gene targeting removes (or otherwise alters) specific DNA sequences in nuclei of cultured cells. mESC are the means by which altered genes are put into whole animals (Figure 7.3). That knockout mice are a stem cell technology is not widely appreciated. In part, this is because results of knockout experiments highlight causal relations between genes and organismal traits, effectively cutting cells out of the explanatory loop. The view that genes are uniquely specific causal agents (see Chapter 6) also suggests that producing a knockout is primarily a matter of altering DNA

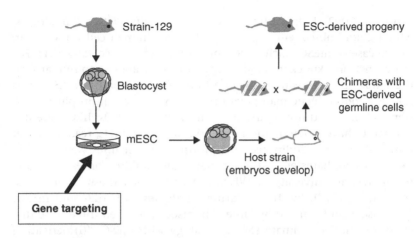

Figure 7.3 Schematic of knockout mouse technique

sequence. However, the stem cell contribution is crucial (Thomas and Capecchi 1987). Gene targeting can be performed on any kind of cultured cell. Resulting cell phenotypes can be investigated in any cell line. But the transfer from cell culture to whole animal is accomplished by ESC, using the blastocyst transplantation method pioneered with mEC. Genetically-altered ESC are added to a normal blastocyst, which is then placed in a mouse uterus to develop into a chimeric mouse. If ESC contribute to the germline of both male and female chimeras, the latter can be mated to produce offspring homozygous for the altered gene (Figure 7.3). The knockout method is restricted to a few mouse strains because ESC can only be made from a few mammalian sources.[150]

Knockout mouse technology is an enormously powerful tool for investigating the effects of specific genes on mammalian physiology and development. For nearly 20 years, the main contribution of stem cells to developmental biology was as part of this fruitful and effective model system.[151] During this period, mESC did not function as model organisms, but as tools mediating between genes and whole animals. So the first embryonic stem cell lines were not models of development. Their epistemic role shifted in the third stage of pluripotency research, which continues today. Once again, the impetus for change was a new embryonic stem cell line.

7.5.3 Human ESC

The first stage of pluripotency research showed that "lessons from the mouse cannot necessarily be extrapolated to human embryogenesis"

(Andrews 2002, 410). Rather, mouse models offer *potential* analogies, which guide subsequent experimentation on human cell lines.[152] But, in the case of mESC, such experimentation lagged (see Table 7.1). For ethical reasons, knockout technology cannot be applied to humans. So there was no impetus to engineer a human counterpart to mESC in its role as a tool for gene manipulation. However, research on pluripotent stem cell lines did not disappear entirely. In the 1990s, hEC were discovered to have differentiation potential similar to mEC, realized under different culture conditions, and pluripotent cell lines were derived from chicken, hamster, and other mammals. Building on earlier work, researchers at University of Wisconsin established rhesus monkey ES cell lines (rhESC), by 'triangulating' analogies with hEC and mESC. The Wisconsin team, led by James Thomson, predicted three "essential characteristics" of primate ESC by analogy with mESC: "(i) derivation from the pre- or peri-implantation embryo, (ii) prolonged undifferentiated proliferation, and (iii) stable developmental potential to form derivatives of all three embryonic germ layers even after prolonged culture" (Thomson et al. 1998, 1145). After success with rhESC, Thomson's group, along with several international collaborators, aimed to produce human-derived stem cell lines with the same three characteristics. Their 1998 success initiated the third, ongoing stage of pluripotency research.

The hESC method is strikingly similar to earlier experiments on pluripotent stem cell lines, differing mainly in having a larger base of models and analogies to build on. The first hESC lines were constructed not from a single model, but via a complex set of comparisons among diverse, partially overlapping models of mammalian development. The basic steps of the method were as for mESC, but specific cell traits selected were those of hEC and rhESC. Analogies with normal human embryos were also important. The initial selection process illustrates the complex relations to pre-existing models that guided construction of hESC. Human blastocysts consist of an outer layer, which can give rise to extra-embryonic tissues, and an inner cell mass (ICM), which can give rise to embryonic tissues. To make hESC, ICM from blastocysts created in vitro are removed and placed on a "feeder layer" of cultured mammalian cells (of mouse origin, in the first experiments). In this environment, some ICM cells divide rapidly, producing clumps of "outgrowth" after 1–2 weeks. Clumps are selected, dissociated, and re-plated on fresh feeders, where new clumps appear in turn. Of these secondary colonies, a few are selected which morphologically resemble undifferentiated cells of early embryos *and* of other ES cell lines (Thomson et al. 1998, 1147, note 6). With each transfer to fresh plates, selection continues. A more detailed comparative profile emerges as cell traits are measured: surface molecules like mESC and rhESC, cell morphology like rhESC, and chromosomal appearance and protein activity like unmanipulated embryos. Similarities to cancer and EC, however, are downplayed, for reasons made clear below (§7.6).

Pluripotency for hESC is demonstrated much as for EC decades before: differentiation in culture to form embryoid bodies and transplantation into adult mice[153] to produce teratomas, the benign counterparts of teratocarcinoma (see §7.5.1). As with EC, the criterion for pluripotency is production of cell types representing each of the three germ layers: ectoderm, mesoderm, and endoderm. The germ layers are a proxy for pluripotency; if there is a match of traits from at least one mature cell type derived from each germ layer, researchers conclude that the cell line is pluripotent.[154] The main contrasts with experiments of the 1970s and 1980s are refinements in cell culture technique that allow researchers to direct hESC differentiation along particular pathways, selectively producing neurons, blood cells, muscle, or other tissue. Like mESC, hESC robustly differentiate into many kinds of tissue, though under slightly different culture conditions and via partly-overlapping molecular mechanisms. hESC were constructed as part of a network of models of early mammalian development, linked by both similarities and contrasts (Figure 7.4).

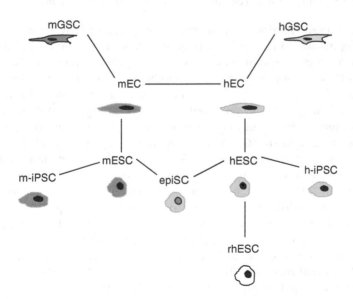

Figure 7.4   Pluripotency network

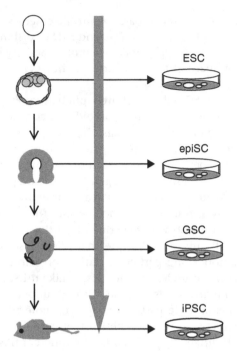

Figure 7.5   Cultured stem cell lines by developmental stage of source

Since 1998, pluripotency research has proceeded by continuously adding models to the network (Figure 7.5). Shortly after the landmark publication of Thomson et al., a second group, led by John Gearhardt at Johns Hopkins University, reported another human embryonic stem cell: the germline stem cell (GSC) (Shamblott et al. 1998). This cell line was derived from primordial germ cells of human embryos and constructed using multiple analogies to other stem cell lines: the mouse counterpart mGSC, hEC, and rhESC. Some efforts have been unsuccessful – most notoriously attempts to produce hESC from human eggs and transplanted nuclei from differentiated cells, which ended in a highly-publicized scandal in 2005.[155] But this disappointment was soon overshadowed by the innovation of iPSC. These products of "cell reprogramming" (see Chapter 6) were originally constructed to resemble ESC, first from mice, then humans. Research on iPSC initially emphasized similarities with ESC; more recent studies examine differences as well, particularly in developmental outcomes. The relation of ESC and iPSC is still unsettled, in part owing to diversity across iPSC lines (an echo of EC variability).

Tracing the origins of pluripotent model organisms illuminates important aspects of pluripotency research today. At each stage, pluripotency research was organized around model systems offering experimental access to mechanisms of development. At each stage, a different research aim predominated, structuring relations among existing experimental systems and guiding construction of new models. In the first stage, the analogy of normal and malignant development guided construction of a network of models, which together revealed similarities and contrasts between teratocarcinoma and early embryonic development. In the second stage, stem cell models of development receded, while genetic manipulation came to the fore. mESC played a significant role, but as a tool rather than a model organism. The latter role returned to prominence after the innovation of hESC in 1998, which ushered in the third stage. Many features of pluripotency research, including the main experimental methods, date back to research on teratocarcinoma and EC in the 1950s. But there are important contrasts. Today, pluripotency research is organized by a web of comparative relations among diverse model organisms: the various pluripotent stem cell lines. As more models are added to the network, new contrasts and similarities are revealed. Similarities with cancer have a quite different significance today. Rather than an overall guiding analogy, they indicate obstacles to a unifying research goal: harnessing stem cell capacities for therapy.

7.6 Model epistemology

Relations among pluripotent stem cell lines are significant in the epistemology of stem cell biology. As model organisms, pluripotent stem cell lines serve as exemplars for new experimental systems, which may or may not become models in their turn. The overall result is a dense network of comparisons among models, which directs and frames ongoing experiments. Much of our current knowledge of stem cells emerges from this interconnected network of model organisms (Figure 7.4).

Within this network, the role of hESC is distinctive in at least three ways. First, hESC lines are foundational in that their establishment unified stem cell biology under its current therapeutic aim. As noted earlier, what is innovative in the landmark paper of Thomson et al. is neither their method, which closely resembled that used for EC, nor the result, pluripotency, a capacity exhibited by many cell lines. What made their achievement a watershed, rather than a mere technical advance, was conceptualizing hESC in terms of an explicit therapeutic goal:

> Progress in basic developmental biology is now extremely rapid; human ES cells will link this progress even more closely to the *prevention and treatment of human disease*... The standardized production of large, purified populations of euploid human cells such as cardiomyocytes and neurons will provide a potentially limitless source of cells for *drug discovery and transplantation therapies*.
>
> (Thomson et al. 1998, 1146–1147; emphasis mine)

These clinical aims unified and galvanized a new biomedical research community: the pluripotency branch of current stem cell research. Human ESC provided a concrete focus for these experimental efforts, transforming embryonic stem cells from a technology to the focus of a new research field. For these reasons, hESC can be considered the foundational model organism for stem cell biology today.

Second, hESC occupy a central place in the network of model organisms. As self-renewing cell lines that can give rise to a wide variety of adult tissues, hESC are intriguing objects of study in their own right. Like classic model organisms, they are constructed to be tractable objects for laboratory study. Cultured ES cells are small, immobile, and continuously accessible to visual inspection and experimental manipulation. hESC lines are selected for continuous, rapid reproduction under controlled conditions. To meet the standard for pluripotency, these lines must also respond to experimental manipulations of their environment by differentiating 'on demand.' Ongoing efforts to vet and standardize hESC lines, such as those by the National Institutes of Health (NIH) Stem Cell Registry and UK Stem Cell Bank, aim to streamline reproducible results across laboratories. Culture media is increasingly well-defined, while risky sources of variation, such as mouse feeder cells, are being eliminated. Like other model organisms, hESC are continuously constructed to be simple, accessible, tractable, prolific, and capable of yielding reproducible results in accordance with researchers' aims. Stringently simplified with respect to other biological functions, hESC are nearly ideal model organisms for studying early stages of human development.

Yet hESC do not 'work alone,' so to speak. As the previous section showed, their main epistemic significance is as part of this network of models, which are used together to reveal molecular mechanisms underlying pluripotency. Since 1998, models of early mammalian development have proliferated rapidly, yielding a dense network of partially-overlapping models. Within this network of models, however, hESC have some special features. Their model organism traits make hESC well-suited to play a key role in construction and assessment of these new models, providing the standard by which they are judged. Advances in pluripotency research since 1998 have been conceptualized primarily in relation to hESC. For example, iPSC are characterized in relation to hESC. Another embryo-derived mouse cell line, epiSC, resembles hESC more than its mouse counterparts. hESC are today positioned at the center of the pluripotency network – a hub of comparison among model stem cell lines.

Third, comparisons among diverse stem cell lines have so far shown hESC to have the most desirable features in light of research goals. Of all the stem cell models constructed to date, hESC exhibit the best combination of wide differentiation potential, low tumor formation, minimal alteration in culture, and fast reproductive rate. These features are desirable for practical reasons; an optimal source of cells for therapy would self-renew, rapidly produce all relevant tissues, and closely resemble normal human cells, with no cancerous, viral, or xenogeneic traits. But they are also epistemically advantageous. A model with these features represents early human development in a simple, accessible way. hESC are, to date, the simplest, easiest to work with, safest, and most understandable stem cells. They are thus *exemplary* models of early human development and so have a special place in the network of models comprising pluripotent stem cell research. A final intriguing feature of pluripotency research is that accomplishment of its therapeutic goal requires effacing the distinction between stem cell models and their primary target, the developing human body. A major hope of regenerative medicine is to transplant stem cells, or their products, into human bodies, to repair injuries and pathological conditions. In this imagined future, stem cell models become their targets – us.

The tri-fold significance of hESC has important implications for science policy. Human ESC research is controversial, primarily because creation of cell lines from embryos disrupts the further development of those embryos. One argument often made against such research is that less morally objectionable methods – "adult stem cells," iPSC, and cloned stem cell lines from unfertilized human eggs, with more doubtless on the horizon – could plausibly deliver comparable clinical benefits. There are two common responses to this anti-hESC position: an argument from ignorance, and an argument from evidence. Neither is very satisfactory. The argument from ignorance is, briefly, that as we do not know where cures will come from, we should pursue all available options to maximize our chance of success. Yamanaka, whose group pioneered iPSC, makes this point very clearly for reprogramming technology:

> At this time it is premature to discuss which method [of cell reprogramming] will ultimately be most appropriate for clinical use. It is important to promote thorough and careful basic research on all the methods. Eventually, such studies could potentially even lead to the development of a new unified technology.
>
> (Yamanaka 2007, 46)

The argument from ignorance rests on two well-established premises: we cannot predict the long-term outcome of present-day research, and a diversified strategy is (*ceteris paribus*) a good bet under uncertainty. But it does not follow that we should use every method at our disposal. Indeed, pursuing *all* possible options is never feasible. Experimental research involves choices and selection at every turn, not systematic exploration of all possibilities. So this argument provides, at best, weak justification for hESC research. Also, it is unsatisfying to mobilize support for embryonic stem cell research by appealing solely to our ignorance, rather than any positive results. So this response is inadequate. The second response – the argument from evidence – is actually a family of arguments. For a given alternative proposed to replace hESC, experimental evidence shows that hESC are more promising, usually with respect to clinical goals. For every alternative to hESC proposed so far, an argument from evidence has been accepted by the stem cell research community: no adequate substitute for hESC has been found. However, the argument must be made anew for each proposed candidate: blood stem cells, the elusive "multipotent adult progenitor cells," iPSC, and doubtless more to come. In each case the same argument is reiterated with slightly different details, and the controversy takes several years to settle. The dialectical process is inefficient and, moreover, puts hESC research continually on the defensive. So, although arguments from evidence do provide strong support for hESC research, another response that streamlines debate and forestalls controversy would be preferable.

The model organism account of hESC and pluripotency research offers such an argument. Eliminating hESC research would not only remove a key model organism from the field, but would also disrupt the epistemic organization of pluripotency research as a whole. Minus the hESC 'hub,' connections among different models comprising the pluripotency network would be drastically reduced. The set of analogies that guide ongoing experiments would thus be impoverished, and the entire pluripotency branch of stem cell biology set back. Connections among mouse, cancer, and iPSC lines would need to be re-examined to coordinate effectively with the overall clinical aim. No existing model is as well-suited to play hESC's central, unifying role. If this entrenched, foundational model were removed, then pluripotency research could well fragment to be subsumed piecemeal by other projects. In this sense, hESC research is indispensable to stem cell biology today.

7.7 Conclusion

Underlying the "indispensable hESC thesis" is the idea that different model organisms in stem cell biology do not substitute for, but complement one another. Rather than competing, model organisms together form a network of complex analogical relations, which collectively offer insights about phenomena less accessible to experiment. These insights do not, however, take the form of direct generalizations. Their role is rather to suggest hypotheses about the latter, which can then be tested experimentally. In biomedicine, at least, model organisms are not surrogates for less experimentally tractable systems. A method with demonstrated efficacy in model systems, such as mice or cultured cells, is not directly generalized to humans. Instead, general knowledge about developmental mechanisms emerges from networks of model organisms, notably stem cell lines, which together reveal robust patterns and causal relations.

A model such as teratocarcinoma-129 can be proposed as a representation of the general causal mechanisms that produce all cell development, normal and pathological. However, any such claim is merely speculative. Rather than a source of generalizations about cancer, say, the teratocarcinoma-129 model is a source of experimental tools and concepts that could be used to understand human cancers. A successful model organism thus serves as an exemplar, thereby influencing research programs on other experimental systems. Models thus generate further models, giving rise to an expanding network of comparative relations across diverse experimental systems. Generalizations about pluripotency and related developmental phenomena emerge from comparisons across diverse models. Such generalizations are therefore relative to the set of models available for comparison.

The next task is to connect this result to the account of mechanistic explanation in Chapters 5 and 6. Both MEx and generalizations about pluripotency result from collaborations that include many experimental systems. The next chapter examines this process, focusing on a watershed case from the field's other branch: tissue-specific stem cell research.

8
Social Experiments

8.1 Introduction

How does robust knowledge of stem cells and underlying developmental mechanisms emerge from a network of model organisms? To answer this question we need to examine the social organization of stem cell experiments. Like many experimental fields, stem cell biology is organized into laboratories, each headed by a Principal Investigator who directs and coordinates the projects of laboratory members. Members include some combination of students (graduate, undergraduate, and medical), post-doctoral researchers, and staff; size and demographics vary widely. Stem cell laboratories exist in diverse institutional arrangements: academic departments, medical schools, private institutes, biotechnology companies, government laboratories, and so on. Most stem cell experiments take place within one laboratory, or a few in close spatial proximity (e.g. Figure 8.1). As discussed in previous chapters, stem cell experiments involve manipulation of model organisms and their parts: organs, tissues, cells, and molecules. In addition, stem cell experiments implicitly model their concrete objects of study with abstract 'models-in-methods.' Results of a stem cell experiment link the two kinds of model.

However, these linkages do not suffice for mechanistic explanation of cell development. Most stem cell laboratories focus on, at most, a few molecules (Bmi-1, c-kit, Notch, and so on), tracing their interactions and roles in exquisite biochemical detail. Explanations of results within a laboratory are thus referred to a 'focal molecule.' If shown to affect development, the latter are often described as 'master regulators' of developmental pathways. But explanations of this kind are inevitably partial, based on experiments that highlight the roles of a few interacting

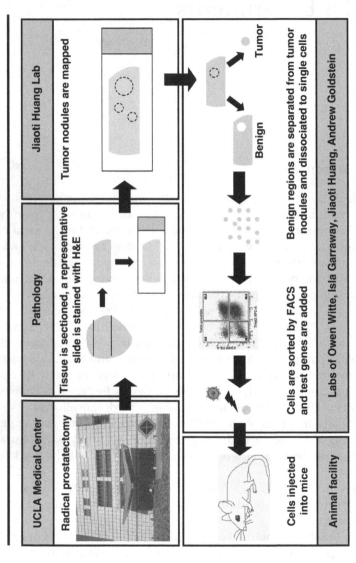

Figure 8.1 A multi-laboratory experiment on tissue-specific stem cells (reprinted with permission of Owen Witte and Andrew Goldstein, UCLA Medical Center)

components of a highly artificial, heavily-engineered, model system. More comprehensive mechanistic explanations (MEx) of cell development emerge from a community of multiple laboratories: an *experimenting community*. Understanding MEx requires a shift in the level of philosophical analysis, from individual scientists or small teams to a more inclusive community.

The first part of this chapter motivates this shift by revealing limitations of the traditional, individualistic view of experiment. I use a case study to argue that stem cell experiments should be conceived not as isolated activities of a single laboratory, but as modes of participation in a wider experimenting community. I then look more closely at the epistemology of such 'social experiments,' using the same case. This community-level analysis reveals ways that experimental methods, social organization, abstract and concrete models intertwine to produce robust knowledge of stem cells. Two are particularly important: first, research teams' methods are at once sources of evidence and modes of community participation; and, second, the structure of abstract models-in-methods reflects community organization. These features exemplify social dimensions at the core of stem cell experiments. My argument for these claims uses a case study from tissue-specific cell research. The case concerns the main exemplar for this branch of stem cell biology: blood, or hematopoietic, stem cells (HSC). These stem cells, the basis of bone marrow transplantation therapy, were the first to achieve wide clinical use and remain the best-understood of all stem cells.[156] This chaper therefore complements the previous, which focused on the origins of pluripotency models. The following sections take a detailed look at models and experiments in tissue-specific stem cell research. The final section discusses the broader philosophical significance of social experiments.

8.2   Experiment and theory

The canonical role of experiment is to provide evidence with which to test theories or hypotheses (Franklin 1986, Mayo 1996). Stem cell experiments indubitably play this role, furnishing evidence to test hypotheses about the behavior of stem cells of various types under various conditions.[157] But this is not their only role. Case studies from physics pave the way to this conclusion. A number of case studies (including Galison 1987, Hacking 1983, and Steinle 2002) reveal a non-canonical, "exploratory" role for experiments in physics. "Exploratory experimentation"[158] differs from theory-testing in aim, principal activities, experimental methods, and instrumentation. Experiments aimed at theory-testing are designed to provide evidence for or against specific theoretical predictions. Their principal activities are acquiring this evidence and securing its reliability using standardized instrumentation arranged to produce a specific effect with minimal variation. In contrast, exploratory experiments aim to establish "low-level regularities" among phenomena and to articulate concepts with which to express these regular patterns clearly and simply. Principal activities involve systematic variation of parameters in order to determine which are essential to producing a regularity of interest. Though these methods are perforce informed by hypotheses about the phenomena under study, these hypotheses are open-ended and provisional. Instrumentation is unstandardized, open to rearrangement, and yields diverse outcomes.

Exploratory experimentation is called for when we lack a stable conceptual background for formulating hypotheses and predictions, such that "the very concepts by which a certain field is treated have been destabilized and become open for revision" (Steinle 2002, 426). These situations arise when research is directed to a new domain of inquiry and in periods of profound theoretical change.[159] Over time, methodologically-diverse exploratory experiments can transform an unstable, indeterminate epistemic situation into a stable, regular system suitable for rigorous theory-testing in a framework of general laws.

This account of exploratory experiments in physics does not straightforwardly extend to stem cell biology. The latter field does not aim to formulate simple, stable regularities, but to articulate detailed MEx of complex systems. Its long-term goal is to produce robust effects in organisms, notably human bodies, and its major successes involve manipulation of living systems in highly specific contexts. *Simple* regularities are not a focus of experimental activity in stem cell biology. Furthermore, there is little indication that stem cell experiments aim to ground general laws or formal theories, as neither figure prominently in stem cell biologists' current discourse and practice. Finally, conceptual innovations do not play a prominent role in stem cell biology. Instead, concepts are taken 'off the shelf' from allied fields such as developmental biology, molecular and cellular biology, biochemistry, cancer biology, biophysics, reproductive medicine, and immunology. There is no shortage of concepts with which to express experimental results, but rather an embarrassment of riches.

The next four sections articulate and defend another view of experiment that complements its traditional role, focusing on the community level and relations among laboratories. Like Steinle's, this community-level account is not intended to replace the traditional view, but to augment it.

I begin with a negative argument. An individualistic, but *prima facie* plausible, account of hypothesis-testing demonstrates the limitations of the traditional view of experiment. The individualistic account focuses on research reports from single laboratories. Hypotheses that are candidates for scientific knowledge appear as conclusions of published research reports. Such reports are not, of course, accurate chronicles or perfect reflections of scientists' experimental practices. Rather, they represent the all-too-rare productive moments of experimental work in the form of an argument to be critiqued by colleagues. A research report is thus a model of experimental practice: a linguistic (and pictorial) description that represents experimental practices in certain respects and to certain degrees, omitting some aspects and emphasizing others for the purpose of justifying the authors' conclusion.[160]

The core of an experimental research report consists of its "Materials and Methods" and "Results" sections, which describe how specific entities and techniques are arranged to produce experimental outcomes. These results are linked to a conclusion by a "Discussion" section, in which alternative hypotheses are rejected and any inconsistencies with previous reports explained. Research reports can thus serve as proxies for experiments that confirm hypotheses. Assuming, for the sake of argument, that the sole epistemic role of experiment is to test hypotheses, then the entire contribution of a stem cell experiment to scientific knowledge may be *read off* its report. I will refer to this as the Proxy View of experiment (PV). PV operationalizes an exclusive interpretation of the traditional account of experiment. Rejecting PV, as the following sections argue we should, does not exclude hypothesis-testing in stem cell biology, but only the assumption that hypothesis-testing exhausts the role of experiment in that field. The alternative I defend, the 'social experiments' view, is not exhaustive either, but contributes to a more complete characterization of stem cell experiments. The case study that follows both rebuts PV (I) and illustrates key features of social experiments (II).

8.3 Blood stem cells (I)

Blood stem cells (hereafter HSC) are found, after embryonic development, in mammalian bone marrow, circulating blood, and umbilical cord. Throughout an animal's life, they replenish the short-lived cells of blood and the immune system (Figure 8.2). HSC were the first non-pathogenic stem cells to be characterized, the first used in routine clinical practice, and remain among the best understood of all stem cell types.

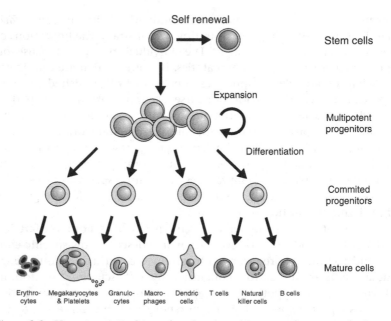

Figure 8.2 Hematopoietic hierarchy showing the currently accepted view of blood cell development. Reprinted from Ema et al. (2010) with permission from Spring Science+Business Media

Mouse HSC was identified and characterized in the 1980s (Doulatov et al. 2012, Kondo 2010, Koretzky and Monroe 2002).[161] Two reports, one published in 1984 and the other four years later, announced the isolation of all and only HSC from mouse bone marrow. The first was by Jan Visser and four colleagues at the Radiobiological Institute in Rijswijk, the Netherlands (Visser et al. 1984). The second was by three members of Irv Weissman's laboratory at Stanford University Medical Center in California, USA (Spangrude et al. 1988). Each described a method with three steps: cell sorting, functional assays, and microscopic observation (Figure 8.3). According to PV, these experiments tested hypotheses of the form 'Mouse bone marrow cells with character values C are HSC.' For each experiment, 'C' refers to details of the methods. The latter need closer examination.

In the first step, sorting, cells extracted from mouse bone marrow were split into distinct subpopulations by size, density, cell cycle status (i.e. dividing or non-dividing), and cell surface molecules. This step used two important experimental innovations of the 1970s: monoclonal antibodies (mAb) and fluorescence-activated cell sorting (FACS).[162] Bone

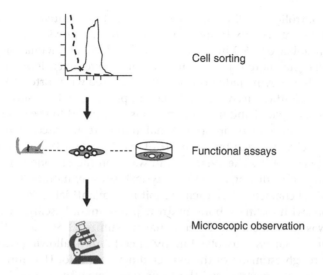

Figure 8.3 Schematic of experimental method for 1984 and 1988 reports of HSC isolation

marrow cells were incubated with specific mAb, which bound to cell surface proteins. The mAb, in turn, were labeled with fluorescent tags. Antibody-labeled cells were then passed through an electrostatic field, differentially deflecting them according to cell size, density, and fluorescence intensity. Cells were thereby sorted into discrete subpopulations based on biophysical characteristics and expression of specific surface molecules. In the second step, each subpopulation was assayed for stem cell capacities. Three kinds of assay were used: in vitro cell culture, spleen colony formation, and radiation rescue. All three measure the capacity to give rise to mature blood cells, but in different ways. In vitro assays measure the capacity to develop into a specific blood cell type in artificial culture. These artificial cultures differed from those of pluripotent SC, being shorter-term, with different molecular factors. But the basic design of the differentiation assays was the same.

Spleen colony formation, the original method for detecting HSC, uses mouse spleen as a medium of "in vivo cell culture" (Till and McCulloch 1961; see §3.4.2). Bone marrow cells, which give rise to the immune system, are highly sensitive to radiation. But irradiated mice can be "rescued" with a transplant of bone marrow cells, which effectively transplants the entire blood and immune system. Spleen colonies are a byproduct: lumpy nodules of $>10^6$ cells that appear on spleens of transplant recipients.[163]

Under controlled conditions, and given several substantive assumptions, the number of "colony-forming units per spleen" provides an estimate of the number of HSC in a given cell population.[164] Radiation rescue is simpler, but more time-consuming: mice were given lethal doses of blood cell-destroying radiation and then injected with sorted bone marrow cells, to either survive for ≥30 days or perish. Survivors showed fully reconstituted blood and immune systems, attributed to the transplants. Finally, cells from each sorted subpopulation were microscopically examined and characterized by methods of cell biology (Bechtel 2006, van Bekkum et al 1971). The overall result was a profile of bone marrow cell subpopulations, distinguished by physical, surface, morphological, and histological characteristics, each exhibiting a quantifiable degree of HSC function, with ordinary bone marrow providing a baseline. The challenge was to find sorting criteria that maximized HSC capacities. But each functional assay involved many uncertainties, allowing derivation of only rough estimates of the expected maximum for HSC function.

Though both reports used the same basic experimental design, they differed in the details. Two differences were particularly important: criteria used for cell sorting and functional assays used to estimate maximum HSC enrichment.[165] Each report's conclusion was stated in terms of the sorting criteria used. The Rijswijk group used cell density, morphological appearance, surface molecules binding wheat germ lectin, and major histocompatibility antigen H-2, and concluded that HSC are high-density, morphologically undifferentiated, wheat germ lectin (WGA)-binding, H-2k$^+$ cells.[166] The Stanford group used surface molecules specific for different blood cell lineages: B and T lymphocytes, monocytes, granulocytes, and one new, specially-made mAb (Sca-1). They concluded that HSC are Thy-1^{lo}B$^-$M$^-$G$^-$Sca-1$^+$ cells. The conclusion that any bone marrow cell population consisted of all and only HSC depended, as noted earlier, on estimates of maximum HSC enrichment. Such estimates were based on parameters of functional assays. However, the basic method involved multiple assays, offering an array of parameters, all involving some uncertainty.

Importantly, the two groups estimated maximum HSC enrichment using parameters from different assays. Visser and colleagues used the traditional method: spleen colony-formation. Weissman and colleagues used radiation rescue of whole animals, the most physiological assay. So the conclusions of the two reports were not directly comparable as alternative hypotheses. Though supported by very similar experiments, they were formulated in different terms and depended on estimates from different functional assays. Unsurprisingly, then, controversy ensued upon publication of the 1988 report. The critical points were summarized in an openly skeptical editorial for *Immunology Today* by prominent HSC researchers,

> But does this represent any advance on previously published data? Are these the real stem cells? Does the report merit its widespread coverage in the newspapers or is this yet another example of indiscriminate glamourization by some sections of the press?
>
> (Lord and Dexter 1988, 376)

8.4 Resolution of controversy

Lord and Dexter's skeptical questions about the Stanford report incisively summarize the HSC controversy. The answers eventually settled upon (though not unanimously) were, respectively, "yes," "no," and "maybe." Accounting for this outcome reveals the limitations of PV. After indicating its problems, I then revisit the HSC case from the social experiments perspective.

8.4.1 Advance

The 1988 report has had considerable impact. It is widely cited,[167] and both first and last authors have co-authored chapters on isolation and characterization of HSC for influential stem cell textbooks (Kraft and Weissman 2005, Spangrude and Slayton 2009). Methods used at Stanford to characterize HSC were quickly extrapolated to other stem cell systems, and now inform the "most rigorous assessment" of tissue-specific stem cell characteristics (Melton and Cowan 2009, xxvii). The 1984 report, though in no way a failure, has had much less impact. So the scientific community has determined that the 1988 report did represent an advance over its predecessor. But the nature of the advance is not obvious. As discussed earlier, both reports used similar methods and results to argue that HSC had been isolated and characterized from mouse bone marrow. Because of the differences of detail already noted, their conclusions do not conflict outright, nor are they directly comparable. So in what way was the Stanford report an *advance*?

Given PV, any epistemic advance by experiment involves hypothesis-testing. Either the 1988 conclusion is better supported by experiment, or the 1988 experiments support a better conclusion, or the advance is non-epistemic. There is no reason to accept the first option.[168] There are no published critiques of the Rijswijk group's results, and retrospective interviews with HSC scientists reveal only praise for the 1984 study (Radetsky 1995, Spangrude 1989, Spangrude pers. comm.[169]). The historical background of the Rijswijk method reveals that its techniques were well-calibrated and conformed to contemporary standards for blood cell research. The 1984 report issued from a long-term research program at the Netherlands' Radiobiological Institute TNO, which began soon after the spleen colony assay was developed. The original protocol was designed to isolate and characterize spleen colony-forming cells, the operationally-defined HSC. The three basic steps – cell sorting, functional HSC assays, and microscopic examination – were all in place before 1970, using, respectively, ultracentrifugation, spleen colony assays, and electron microscopy. When the co-authors of the 1984 report began to assemble at the Institute, all three technologies had been used there to purify and characterize blood cells for over a decade, and standards for distinguishing reliable experimental results from instrumental artifacts were well-established.

Throughout the 1970s, additions to the basic protocol increased its specificity and provided additional lines of evidence. mAb technology offered new ways to distinguish morphologically similar blood cells by binding surface proteins with exquisite specificity. The Institute acquired one of the first commercially-available FACS apparatuses and began collaborating with other Netherlands institutes on various "cytofluorometric" projects (Keating and Cambrosio 2003; Morselt et al 1979). Jan Visser, in particular, saw the new instrument as a tool for characterizing blood cells in rigorous physical and chemical terms. He and various colleagues spent years optimizing FACS to purify and characterize blood cells. In these experiments, light scatter (indicating cell size and complexity) and surface molecule expression were correlated with HSC function in increasingly narrowly-defined cell populations. The discovery that HSC function correlated with WGA-binding and H-2k expression emerged from a long and painstakingly optimized series of FACS experiments. In the years following the 1988 controversy, Visser and colleagues wrote a number of well-received technical articles on the use of FACS in HSC research, including a chapter on "Analysis and sorting of hematopoietic stem cells from mouse bone marrow" in *Methods in Cell Biology* (Visser and de Vries 1994). Overall, the Rijswijk group's 1984 conclusion was supported by multiple functional assays using well-calibrated techniques. Though their argument included several questionable assumptions, the Stanford report was no better off.

The 1988 report was also the culmination of many years of research, but along rather different lines. In the early 1970s, when the Stanford

HSC project began, the immune system was divided into three main lineages of blood cells implicated in immune reactions: B, T, and myelo-erythroid (M-E) cells. Each lineage develops from a then-uncharacterized "progenitor" cell. Members of the Weissman laboratory aimed to characterize these progenitors, working 'backwards' from later to earlier developmental stages. So the Stanford group's HSC research began as three parallel projects, with different in vitro assays to detect progenitor cells in each of the three lineages. Each experimental assay of blood cell development was combined with mAb-FACS technology. Here, the Weissman group had a distinct advantage. FACS technology was developed at Stanford. The prototype machine became operational in the early 1970s, and was made available to members of "the FACS shared users group" in 1974.[170] The Weissman laboratory was thus among the first to use FACS and the very first to use it to analyze early stages of blood cell development. When the group began to seek HSC, its members had been using FACS for over a decade, with the advantage of unmatched on-site technical expertise. Their 1988 report built on nearly two decades of technical refinement. The question remains, however: was their conclusion a distinctive experimental contribution, or did it merely replicate Visser and colleagues' earlier result?

The controversy was not a priority dispute, as there was never a question that the Rijswijk group's report came first. The problem was not how to allocate credit among two groups with the same result, but whether their results were really the same. The quality of evidential support, at any rate, does not distinguish them.

8.4.2 Not the real stem cells

The other epistemic possibility, given PV, is that the Stanford experiments supported a better conclusion; that is, the 1988 hypothesis stood up better to subsequent tests than its 1984 counterpart. Many scientific controversies are resolved in this way, as experimental evidence gradually accretes to decisively favor one hypothesis over another. But the HSC controversy did not play out this way. Instead, further experiments, by other groups as well as members of the Rijswijk and Stanford teams, disconfirmed *both* conclusions. Post-1988 experiments showed that bone marrow cell populations isolated using either method were not pure HSC, but heterogeneous mixtures of HSC and more differentiated blood progenitor cells (Spangrude 1989, Spangrude pers. comm., Visser and de Vries 1988). Within a year, there was broad (though not unanimous) consensus that *neither* report had characterized "the real stem cells."[171] So the Stanford report's success was not in virtue of its hypothesis being better-confirmed by experimental tests than that of the Rijswijk report. Both hypotheses were equally well-supported by experimental methods and results reported at the time, and equally poorly by subsequent experiments. There is no reason to consider the 1988 conclusion a better hypothesis than its predecessor.

8.4.3   Indiscriminate glamourization?

The only remaining option, given PV, is that the Stanford group improved on the Rijswijk report in some non-epistemic way. The later report has clearly received more attention than its Rijswijk counterpart, both in the scientific community and the popular press. For my purpose here, it is the scientific community's reaction that is important. Was the outcome of HSC controversy due to "indiscriminate glamourization"? That is, were the reports' differential impacts the result of differential attention and prestige, and not of evidential differences based on experiment? There is reason to think audience and prestige were involved to some extent. The 1984 report was published in the *Journal of Experimental Medicine* and the 1988 paper in *Science*. Weissman called a press conference to announce the 1988 result, which prompted Lord and Dexter's withering remarks about "widespread coverage in the newspapers" and "indiscriminate glamourization by some sections of the press." Institutional and national contexts could also be implicated, as the European research institute and US university medical school contrast in many ways.

The problem with this diagnosis is that PV offers no further insight into the case. If the role of experiment is just to test hypotheses of co-authorial research teams, then all that can be said of this "turning point" in stem cell research is that the outcome was "indiscriminate;" not a justified experimental success. The 1984 and 1988 reports were on par with respect to experimental support, yet had strikingly different impacts on HSC research and stem cell biology more generally. PV cannot account for the contrast. This is a serious lacuna, as the HSC episode is neither isolated nor atypical. Its ramifications extend beyond blood cells, influencing current standards for stem cell and cancer research more generally (see §8.7). The features that problematize PV are even more widespread: rapid obsolescence of authorial conclusions, high turnover of methods and technologies, "indiscriminate" media attention. A less restrictive view of experiment is needed. Rejecting PV allows us to understand how scientific communities, not only individual scientists or small teams, contribute to experimental research.

8.5 Blood stem cells (II)

Revisiting the HSC episode as a case study of 'social experiments' provides a more satisfactory account of its outcome. The key is to conceive of experiments as involving communities of multiple research teams. Individual research teams' methods and results are then seen as modes of participation in a wider experimenting community. The Rijswijk and Stanford reports emerged from distinct experimenting communities. On this view, the HSC case is not a 'duel' between conflicting hypotheses, with one decisive winner, but a 'merger' of two experimenting communities, which produced the explanatory framework HSC researchers have used ever since. This section examines contrasts in the Rijswijk and Stanford methods as modes of participation in their respective communities. This approach reveals a vindicating structural correspondence: the abstract models implicit in the Rijswijk and Stanford experiments mirror wider community organization.

8.5.1 Division of labor

The Rijswijk group was part of a wider research community of blood cell experts (hematologists), the linchpin of which was the spleen colony assay (see §3.4.2). Developed by a small group at Toronto's Ontario Cancer Center, the assay was rapidly adopted and modified by many other groups. Centers of HSC research sprang up in Asia, Europe, and the United States. Meanwhile, researchers at Toronto continued to refine the original spleen colony assay. So the hematological HSC community developed in a manner roughly analogous to spleen colonies, its raison d'être; from a single point of origin, research expanded and methods diversified. However, 'differentiation' of the hematological HSC community contrasted with its biological counterpart in one key respect. Diversification of HSC methods led to fragmentation, rather than a cohesive community. The results of spleen colony variations did not 'add up to' a coherent picture of blood cell development. Melbourne researchers focused on in vitro regulation of cell development, adding and removing specific growth factors from culture media and observing effects on M-E cells. A group at Manchester's Paterson Laboratories pioneered the concept of a stem cell niche, attempting to replicate bone marrow cell-cell interactions in vitro. Various groups in the eastern and mid-western United States studied intermediate 'progenitor' stages between HSC and mature blood cells. The hematological HSC community expanded and diversified its methods, but their work remained patchy and fragmented until well into the 1980s.

Hematologists in the 1960s and 1970s were confronted by complex processes, which could not be comprehensively analyzed with techniques available at the time. Their community's division of labor was therefore unsystematic, with different groups studying different aspects of a largely-uncharacterized process. Within this diffuse experimenting community, researchers at the Rijswijk Institute took the part of isolating and characterizing HSC. The centerpiece of their method was a relatively unmodified version of the spleen colony assay, which defined HSC as cells capable of colony formation (§8.4.1). Visser and colleagues aimed to localize HSC capacities to a "well-defined" subpopulation of bone marrow cells by judicious "combination of parameters" (Visser et al. 1980). "Well-defined parameters" were biochemical, physical, and morphological properties of cells that were understood independently of HSC capacities, having been validated by other, unrelated experiments on mammalian cells. The Rijswijk group scoured the literature for properties that might distinguish HSC from other bone marrow cells: size, density, morphology, purified surface molecules, drug-sensitivity, and so on. Their criteria for cell sorting were gleaned from wider hematology and cell biology communities.

As new potential markers were characterized by other cell biologists, Visser and colleagues continually updated their method, sorting bone marrow cells into increasingly finer-grained subpopulations, of which one was increasingly "enriched" in HSC capacities: 10- to 40-fold, then 60- to 100-fold, and finally 120- to 200-fold (Visser and de Vries 1988). The result was an increasingly detailed 'cell profile' correlated with HSC function. The latter was measured in terms of spleen colony formation, corroborated by radiation rescue and in vitro assays for specific blood cell lineages (see §8.3). By 1988, however, identification of HSC with spleen colony-forming cells had been supplanted by a more complex, three-level hierarchy of blood cell development: HSC, various lineage-committed progenitor cells, and fully differentiated blood cells (Figure 8.2). Depending on experimental conditions, spleen colony-forming cells could be HSC, progenitor cells, or a mixture of the two. So spleen colony formation became an ambiguous indicator of HSC. Radiation rescue, which unambiguously indicated full immune and blood cell reconstitution, replaced spleen colony formation as the standard for measuring HSC.

Because the hematology community was widely dispersed and fragmented, this shift in standards was gradual and uneven. The Toronto group described heterogeneity among spleen colony-forming cells as early as 1963. But the ambiguity of spleen colony formation was not widely recognized until the early 1980s. Differences of aim and method hampered comparison of results across groups, making standards for measuring HSC patchy as well. However, between 1968 and 1988 opportunities for interaction among hematologists increased, with new societies, annual meetings, and dedicated journals countering community fragmentation.[172] As the hematological HSC community became more integrated, the association of HSC with spleen colony-forming cells became more tenuous. Because the Rijswijk group's method interfaced with the wider community primarily in regard to sorting criteria, they retained the spleen colony standard rather longer than most. Their 1988 'cell profile' included size, density, morphology, surface molecules, and drug sensitivity – all features independent of HSC and associated developmental pathways. Though Visser and colleagues made steady progress by their own lights, the correlation between their well-defined cellular properties and HSC capacities was undermined by experiments elsewhere. By 1988, the Rijswijk group recognized these shifting standards and the need to "redefine our concepts" and "further define the stem cell" (Visser and de Vries 1988, 382).

8.5.2 Center of collaboration

The Stanford group belonged to a quite different community, focused on cells with immune function rather than blood cells per se. The 1970s and 1980s were a highly productive period for cellular immunology, and Stanford's Medical Center became a hub of international research in the area. There were actually two distinct immunology communities at Stanford at the time: one biochemical, focused on Ab–Ag binding and complement reactions; the other investigating immune cells and immunogenetics.[173] The latter was knit together by frequent lectures and seminars by prominent visiting immunologists, regular Bay Area Immunology meetings, and on-campus discussion forums that included medical students, clinicians, faculty, and research fellows. Participants in these discussions formed the core of the "shared FACS users group," whose members paid dues to support the new facility and, in return, were allowed to use the prototype machine (see §8.4.1).

From the early 1960s, Irv Weissman participated in Stanford's cellular immunology community; first as a medical student, then as a research fellow in Radiology, and finally as a professor. His training coincided with a watershed for the field: elucidation of the role of lymphocytes (B and T cells) in immune function by James Gowans and his colleagues at Oxford. After visiting Gowans' laboratory in 1967, Weissman extended that group's "cell-tracking" approach to immune cell development.[174] The Weissman laboratory grew rapidly through active recruitment of students, technicians, and post-doctoral fellows. Membership was diverse; Weissman traveled internationally to recruit young immunologists, molecular biologists, developmental biologists, and clinical researchers, and built a worldwide network of collaborators at other centers for cellular immunology. The Stanford laboratory thus became a site for interaction of many different methods, concepts, and standards. Disagreement was tolerated and members had considerable autonomy over their research projects. By the late 1970s, the Weissman laboratory was large and diverse enough to support multiple subgroups working on different aspects of blood cell development, using model systems that ranged from yeast to inbred mice, colonial ascidians to human cell cultures. One methodological norm was widely shared, however. Single-cell measurement was the laboratory standard for in vitro assays, cell sorting, and in vivo radiation rescue. Working to this standard, members of the Weissman laboratory collaborated with one another and with other immunology groups, devising experimental methods to work out relations of different immune cells to one another and to other blood cells.

Every step of the 1988 method was a product of collaboration. Sorting criteria were the cell surface molecules associated with particular immune cell lineages: B cells, T cells, and others. Each major lineage was studied by a different group within the Weissman laboratory, with membership shifting as researchers joined and departed.[175] By the mid-1980s, each subgroup had refined a method to detect single cell progenitors of one of the three main immune cell lineages. But their methods were not yet linked in a coordinated search for HSC. The "turning point" was a collaboration between post-doctoral fellow Christa Müller-Sieburg and medical student Cheryl Whitlock, who discovered a B cell lineage marker (B220), the *absence* of which correlated to enrichment in all three progenitors.[176] Their result suggested that different immune cell lineages share a single source: the elusive HSC. Previously-independent lines of research suddenly converged. From then on, the Weissman group used an integrated 'negative selection strategy' to seek HSC, combining all three assays.

By 1986, most of the components of the 1988 experiment were in place. Though the basic steps resembled those of the Visser group, the pace was accelerated by concentration of resources within a single laboratory. Members shared information rapidly and had continuous opportunities for collaborative interaction. Microscopy was de-emphasized, as mAb for specific cell surface molecules allowed finer separation of blood cell types and stages. FACS was integrated while that technology was in the prototype stage. Functional assays for each major blood cell lineage were incorporated along with cell sorting criteria. Radiation rescue provided a comprehensive measure of differentiation in all the blood cell lineages. The final component of the 1988 method was the result of a collaboration with West German immunologists, which yielded a new mAb for sorting bone marrow cells: "stem cell antigen 1" (Sca-1). From the 1960s onward, the Weissman laboratory served as a continuous center of collaboration, such that its 1988 method was both efficiently assembled and thoroughly integrated with the leading edge of cellular immunology. As with the hematologists, the community resembled its subject matter: developmental connections among diverse blood cell types were investigated by forming collaborative connections among research teams. Coordination of different experimental approaches was aided by a shared methodological norm (single-cell assays) and physiological test for blood cell differentiation (radiation rescue).

8.6 Contrast in experimenting communities

The Rijswijk and Stanford methods are composed of the same basic steps, yield similar results, and are equally well (or badly) confirmed by experiment. Therefore, PV does not account for their differential impact. But the social experiments approach does. Moreover, the social account of the HSC controversy reveals how explanatory frameworks for MEx are established at the community level. Table 8.1 summarizes the relevant contrasts between the 1984 and 1988 experiments. The Stanford group was part of a large laboratory, with a leading role in the international immunology community. This laboratory served as a center of continuous, cumulative collaboration aimed at understanding the inter-related pathways of immune cell development. The 1988 method was thoroughly integrated into this collaborative network, characterizing HSC as a point

Table 8.1 Comparison of experimenting communities of the 1984 and 1988 reports

	Hematologists	Weissman laboratory
Shared core	Spleen colony assay	Immune cell development
Standard	Well-defined cell properties; colony-formation	Single cell tracking; radiation rescue
Organization	Widely-distributed, fragmented, episodic interaction	Continuous collaboration at a center

of convergence among diverse projects. The Rijswijk group, in contrast, belonged to a more methodologically-fragmented community. Their HSC method interfaced with this community at two distinct points: a shared origin in the spleen colony assay, and choice of well-defined sorting criteria for bone marrow cells. Because the latter were related to HSC and one another only via correlation with spleen colony formation, the Rijswijk group's results receded in significance when standards shifted in the wider community.

These differences are hinted at in the reports' conclusions. The conclusion of the 1984 report is formulated in terms of diverse cell traits (high-density, morphologically undifferentiated, WGA-binding, H-2k$^+$), while its 1988 counterpart is articulated in terms of molecular markers for different blood cell lineages (Thy-1^{lo}B$^-$M$^-$G$^-$Sca-1$^+$). But the contrast is clearly brought out only when methods supporting each conclusion are considered in the relevant community context. Each method implicitly modeled HSC in a way that reflected the organization of its experimenting community. The Rijswijk model was a cytophysical profile of colony-forming cells. The profile was a collage of morphological, physical, and cell-surface properties, characterized independently of one another and of HSC capacities. The latter were measured by three kinds of functional assay, with the spleen colony assay providing the standard for quantitative estimates. In particular, correlation of blood cell traits with HSC capacities was calculated using parameters of the spleen colony assay. This correlation, established by the spleen colony assay, was the crux of the model's structure. The structure also reflected that of the hematological HSC community, which originated with the spleen colony assay and then fragmented into diverse projects. Because of this structure, the Rijswijk model was vulnerable to disarticulation. When the spleen colony assay was replaced by another standard, that model was no longer useful. Accordingly, the 1984 conclusion receded in significance.

In contrast, the Stanford model represented relations of HSC to other blood cell types in a branching tree of cell development. Like its counterpart, this model reflected the social organization of its experimenting community. Diverse research teams, within and outside the Weissman laboratory, characterized the phenotype and function of different immune cell types. Developmental connections between blood cells of different types and stages were worked out by a network of collaborative experiments. The Stanford model was constructed to track connections among cell types, correlating phenotype to function at every stage. By focusing on single cells in a coordinated system of bioassays, the Weissman group characterized HSC as the unique stem of a unified model of blood cell development, a point of convergence of multiple lines of research. Like the community that gave rise to it, the model has a tightly integrated structure, and connections between different components are given a prominent place. Accordingly, the model is robust to new experimental data and changing standards for estimating HSC function. Though its specific details were quickly superseded, the basic framework was not. This general framework, cell lineage markers delineating developmental pathways, served as the starting point for increasingly detailed mechanistic explanations of blood and immune cell development (Figure 8.4).

Furthermore, the Weissman group's model conformed to the experimental standards of the hematologists' community. By 1988, radiation rescue had largely replaced colony formation as the standard for operationally defining HSC, and the need for single-cell assays to avoid conflating stem and progenitor cells was widely recognized. Multiple, branching developmental pathways fit with the emerging consensus view of a three-level hierarchy of HSC, committed progenitors, and mature blood cells. These similarities allowed the two experimenting communities to merge. Before 1988, the HSC community had been relatively isolated from immunology, while the Weissman group's cellular developmental approach had focused on immune cell lineages rather than blood cells in general. The 1988 controversy brought them into contact, after which the robust Stanford model guided new experiments in the expanded HSC community. So the Rijswijk and Stanford experiments had very different epistemic impacts. Though both made a contribution, the latter's was more lasting, supplying the basic framework for MEx of blood cell development, and helping to establish the experimenting community that continues to fill out mechanistic details.

8.7 Further examples

The HSC model and method were quickly extrapolated to other experimental systems: from mouse to human HSC; from blood to brain, gut, skin, muscle, liver, pancreas, and the enteric nervous system; from normal to cancerous development, including leukemia, colon cancer, breast cancer, and prostate cancer. In this way, the Stanford group's approach, comprising both method and implicit model, is exemplary for tissue-specific stem cell research. So the results of this case study shed light on an entire branch of stem cell research today. Concepts, models, and methods honed by experiments on HSC have been extended to many other stem cell systems. The current "gold standard" for identifying tissue-specific stem cells is a single-cell transplant leading to long-term

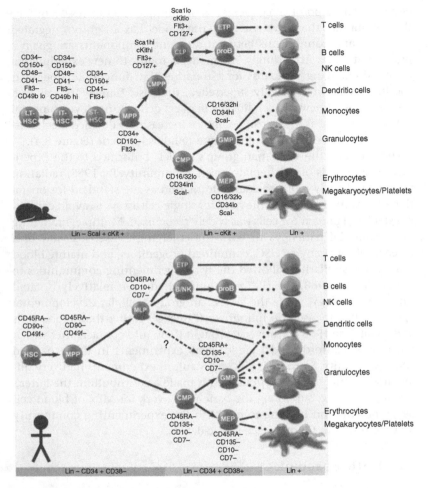

Figure 8.4 Blood cell developmental hierarchy, including molecular 'markers.' Reprinted from Doulatov et al. (2012) with permission from Elsevier Press

reconstitution of the relevant tissue in vivo (Melton and Cowan 2009, xxvii). Social experiments have also been important at other moments of tissue-specific stem cell research. Two important examples illustrate their pervasiveness in the field.

8.7.1 Spleen colonies

The spleen colony assay, which provided the first experimental demonstration of stem cells in mammals, originated as a cross-fertilization of radiation biology, cancer research, and molecular biology. The method was invented at the Ontario Cancer Center in Toronto by two researchers, James Till and Ernest McCulloch, who combined expertise in biophysics and hematology to develop a quantitative assay measuring the effects of radiation on different kinds of blood cells. The role of spleen in blood cell development and immunity was of great interest to radiation biologists from the 1950s onward. Bone marrow transplantation therapy was then in its early stages, so the potential for clinical relevance was clear. The spleen colony assay thus emerged at the interface of several experimenting communities.

Two features distinguished Till and McCulloch's collaboration from other groups working at the interface of radiation and cancer biology. First was their close coordination of statistical argument and concrete experiment, an effective integration of complementary areas of expertise. Second was a key analogy with molecular biology. Till and McCulloch recognized that an irradiated mouse was analogous to a tissue culture environment, and spleen colonies to bacterial colonies:

> ... the irradiated mouse may be considered as providing the receptacle, the medium, and control of temperature, pH, and humidity required for the cultivation of marrow cells ... [Spleen colonies offer] a direct method of assay for these [normal mouse bone marrow] cells with a single-cell technique
>
> (1961, 220, 213)

Unlike bacterial colonies, however, spleen colonies contained multiple types of blood cell, including morphologically unspecialized cells that could give rise to more spleen colonies. Based on these features (extensive proliferation, differentiation into multiple blood cell types, and self-renewal), colony-forming cells were identified as blood stem cells in mice. Though innovated by a single group, the spleen colony assay quickly became the focus of a new experimenting community, with centers in Toronto, Melbourne, Rijswijk, Manchester, and the eastern United States, as described earlier.

8.7.2 Cancer stem cells

Cellular immunology and cancer research merge in studies of leukemia, a condition in which immune and cancer cells are one and the same. In the 1990s, cancer researchers hypothesized that leukemias arise from mutations that alter regulation of self-renewal in HSC, leading to overproduction of undifferentiated blood cells. This hypothesis was supported by multiple converging lines of research. First, experiments showed that different leukemia cell populations varied with respect to self-renewal. Studies of HSC self-renewal were then modeled on these leukemia studies and mechanistic explanations (still works-in-progress) constructed. Finally, the "leukemia stem cell" hypothesis united the two strands to explain the etiology of blood cell cancer. The cancer stem cell hypothesis extends this idea from leukemia to solid tumors, and to cancers in general.

The core of the cancer stem cell hypothesis is an analogy between tumor and organ development: both, researchers speculate, are maintained by self-renewing stem cells that constitute a small subpopulation of cells in the tumor or organ as a whole. The hypothesis is supported by evidence that tumor cells are heterogeneous with respect to self-renewal, with "tumorigenicity" restricted to a small subset of putative cancer stem cells. This idea has important clinical implications. If the analogy holds, then most cells in most tumors are comparatively benign. Cancer therapy should then target only those cells capable of long-term self-renewal: the cancer stem cells, analogous to HSC in bone marrow. The relation to normal development would go beyond analogy if the same mechanisms regulate normal and malignant processes of self-renewal and differentiation. The cancer stem cell hypothesis connects the communities of developmental and cancer biology, focusing both on the problem of identifying and characterizing an elusive population of self-renewing pluripotent cells. Furthermore, the HSC model and method exemplify a solution to this very problem. The negative selection strategy honed in the Weissman group's search supplies the method and evidential standards for identifying and characterizing cancer stem cells: cycles of FACS-mAb experiments, sorting heterogeneous tumor cell populations into increasingly homogeneous 'subsets.'[177] Testing the cancer stem cell hypothesis requires social adjustments. In particular, new interactions are needed, between clinical oncologists and researchers investigating various stem cell types, to produce integrated experiments like that depicted in Figure 8.1. In the cancer stem cell case, social experiments and hypothesis-testing converge.

8.8   Conclusion

The first part of this chapter uses an individualistic account of hypothesis-testing (PV) to bring out important aspects of the role of experiment in stem cell biology. This is not to deny that hypothesis-testing plays a role in stem cell biology. But it is not the whole story. To understand how experiment contributes to knowledge of stem cells, we need to consider research teams' methods as modes of participation in a wider experimenting community, as well as sources of evidence to test hypotheses. This section discusses some philosophical implications of this approach. The structure of abstract models involved in the HSC controversy reflects the organization of their respective communities. This case illustrates how details of experimental methods, social organization of experimenting communities, and structures of abstract models are brought together to generate MEx of cell development. The HSC model and method provided the basic framework, details of which are still being filled out (see Figure 8.4). This approach was quickly extrapolated to other organs and tissues, normal and pathological.

HSC is thus a model system: an object of study in its own right and a basis for comparison and extrapolation to other systems. Insights in other systems in turn 'feed back' to further refine the HSC model, elaborating the simple three-level hierarchy to an intricately-tiered structure of lineages. Like its pluripotency-focused counterpart, the epistemic community of tissue-specific stem cell research has grown as a reticulated network of experimental systems (see Chapter 7). In both major branches of stem cell biology, experimental success has involved coordinating diverse model systems so their results can be compared and integrated. As with pluripotent stem cells, tissue-specific stem cell research is organized as an array of models, each understood in terms of developmental pathways associated with an organ or tissue: blood, skin, brain, muscle, and so on. These pathways, and underlying molecular mechanisms, are worked out by analogies among the various tissue-specific model systems. The better-understood systems guide investigation of others by (for example) suggesting key molecules to track or cell signaling interactions to look for. Within this network of models, HSC occupy a central place.

This shared pattern, and the history of both branches, suggests that the next experimental milestone will be a comprehensive account of cell development that includes both tissue-specific and pluripotent stem cells. Research along these lines is currently underway in studies of endodermal, neural crest, and mesodermal stem cells, as well as efforts to pinpoint the embryonic origin (or origins) of blood stem cells. The entire stem cell community would be advanced by a 'merger' between research on directed differentiation of pluripotent stem cells, on the one hand, and efforts to trace the common origins of tissue-specific stem cell populations, on the other. If the past pattern of success continues, then the two branches of stem cell research should converge. Another progressive outcome would be closer ties between developmental and cancer biology. Merging these fields into a single experimenting community will make possible a more comprehensive account of cell development. Philosophers can facilitate these outcomes by making explicit the means and ends of social experiments.

The epistemic contributions of social experiments, exemplified by the branching hierarchy of blood cell development and its molecular details, are here conceived as abstract models. These models are community achievements, rather than the work of individual scientists or a single team. Their relation to *theories* depends on what is meant by the latter. Social experiments' models perform several traditional functions of theory, including accurate representation, derivation of fruitful hypotheses, and explanatory unification. If 'theory' is understood in purely functional terms, as that which represents, explains, and yields predictions, then the models resulting from social experiments are theories. But if a theory is defined as a linguistic entity or abstract structure, then social experiments' models (in stem cell biology, at any rate) are not theories. Unlike sciences such as physics, economics, and evolutionary biology, stem cell biology does not include an explicit formal framework of axioms or laws. Finally, if theories are conceived as distinct from experiment, requiring mediation by models to connect with experimental phenomena, then social experiments' models are not theories. The models of experimenting communities are closely connected to concrete experimental methods. There is no more abstract theoretical domain 'above' them, which structures stem cell biology today.[178]

Finally, the social experiments account, as illustrated in the HSC case study, clarifies the significance of diverse experimental methods. The importance of independent lines of evidence for confirming theories and hypotheses has been much discussed. But the importance of converging *methods* has not. In successful social experiments, it is not results of experiments that converge, but models implicit in experimental methods. Scientists can then combine these methods into a single, more inclusive experimental design. This unifying strategy is demonstrated by progress in tissue-specific stem cell research and plausibly extends to the field as a whole. The basic idea is that general or comprehensive explanations result from 'crosslinking' results from different laboratories to produce an inclusive network. The process is gradual and decentralized, and the resulting explanations are not simple. In this way MEx are produced by a diverse experimenting community.

Part III

9
Integrating Stem Cell and Systems Biology

9.1 Introduction

Both stem cell and systems biology are newly prominent approaches to studying cell development. In stem cell biology, the primary method is experimental manipulation of concrete cells and tissues, while systems biology prioritizes mathematical modeling at least as highly as experiment. The two fields are ripe for interaction, but collaborations are just beginning. This chapter builds on results of Part II, examining the significance of integration with systems biology for mechanistic explanations (MEx) of cell development. I argue that both approaches are necessary for such MEx and propose a unifying model that visualizes their interdependent roles: Waddington's landscape, introduced in Chapter 6. I begin by sketching the main ideas and commitments of systems biology.

9.2 Systems biology: overview[179]

Systems biology is a loose assemblage of research programs unified by the aim of "fundamental, comprehensive and systematic understanding of life," as well as shared commitments regarding how to achieve this goal (O'Malley and Dupré 2005, 1271). These core commitments are to systematic data collection, mathematical modeling, and integration. The last is actually a family of commitments, to explanatory, evidential, methodological, epistemic, and social integration of, respectively, levels of biological organization; multiple datasets; experimentation, mathematical modeling and computer simulation; "discovery- and hypothesis-driven" modes of inquiry; and different scientific fields and disciplines. Some of these integrative commitments are also characteristic of stem cell research and experimental biology more generally; these are discussed in other chapters. Those that are not shared are potential obstacles to integration with stem cell biology. It is these distinctive commitments of systems biology that are of interest here.

The most obvious of these is commitment to mathematical modeling as means to "system-level understanding" (Aggarwal and Lee 2003, 175). Explanations in experimental biology are predominantly qualitative, describing how interacting components jointly constitute an overall working mechanism (see Chapter 5). Systems models, in contrast, eschew qualitative descriptions, instead representing patterns of dependence among molecular traits, such as concentration or binding affinity, as sets of differential equations. As these equations often resist analytic solution, computer simulations are widely used to derive results and elicit predictions.[180] Different 'schools' of systems biology prefer different mathematical and computational tools. A deeper division within the field concerns theories imported from physics, engineering, and computer science. Some systems biologists claim that genuine understanding of biological phenomena requires "fundamental design principles" incorporating theoretical concepts from these other fields, such as robustness, noise, and modularity.[181] Others put less emphasis on fundamental theories, adopting a more pragmatic approach in which mathematical modeling remains central. These two groups of systems biologists, termed "systems-theoretic" and "pragmatic," respectively, thus have rather different epistemic aims.[182] The former, but not the latter, seek to put experimental biology (including stem cell research) on a new theoretical foundation.

Twentieth-century attempts to mathematicize or import physical theory into biology were largely unsuccessful (Keller 2002). Systems biology today is distinguished from these precursors by another commitment, to systematic, or "global," experimental data. New high-throughput technologies ("omics") allow simultaneous measurement of thousands of molecules and their interactions, putting the aim of comprehensive datasets, the complete measurements of all a system's components, within reach – at least for certain kinds of components. Genomics applies the systematic approach to sequences of DNA, functional genomics to products of transcription (mRNA), proteomics to protein sequences, and so on. Omics technologies underwrite the aspiration to comprehensively measure a system's components, both in number and kind, producing a complete, quantitative 'parts list.' Because biological systems change over time, measurements must be made repeatedly over significant intervals; a truly comprehensive parts list is temporally ordered. In systems biology, measurements aimed at this comprehensive ideal replace

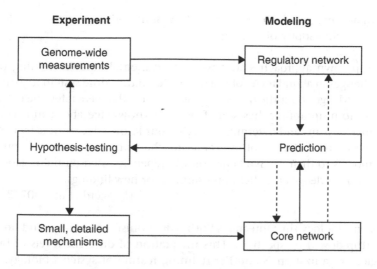

Figure 9.1 Schematic method of systems biology, showing interplay of modeling and experimentation (solid arrows) and of top-down and bottom-up systems approaches (dotted arrows)

the traditional experimental strategy of focusing on one or a few components hypothesized to play a significant role in determining a system's overall behavior. Experiments that use high-throughput methods are therefore often contrasted with "hypothesis-driven" research.[183]

A third core commitment is to an integrated method, including both mathematical modeling and experiment. The method of systems biology is often reflexively visualized as a dynamic system; a feedback loop depicting iterative cycles of modeling and experimental test (Figure 9.1). In this pleasingly coherent sketch, experimental data are used to construct models of molecular mechanisms, which are tested by constructing predictive models or running simulations. Results are then compared with those of analogous experiments on biological systems. Simple, incomplete models of biological phenomena can thus be progressively elaborated and improved, until their outputs closely match those of concrete biological mechanisms.

Figure 9.1 also indicates another integrative aspect of systems biology: "top-down" and "bottom-up" approaches to modeling biological phenomena. Top-down approaches use computational methods to infer details of molecular mechanisms from "global" datasets, while bottom-up approaches represent smaller, "core" mechanisms as systems of equations. The two modeling approaches complement one another. Furthermore, both are integrated with experiment. As summarized in a recent collection on philosophy of systems biology:

> Top-down approaches start with experimental data concerning the changes in abundancies of the molecules in the entire system to extract knowledge about the molecular mechanisms that generated the data... Bottom-up approaches start from the knowledge about molecular mechanisms and determine whether our knowledge suffices by comparing the behavior of molecular mechanisms in detailed computer models to their in vivo behavior. Discrepancies then pinpoint gaps in our knowledge and offer opportunities for new findings.
>
> (Boogerd et al. 2007, 322)

Systems biology thus integrates both reductionist (bottom-up) and holistic (top-down) perspectives. This integration of different levels of biological organization is another defining feature of systems biology. In this respect, however, it resembles experimental biology, which aims to construct interlevel MEx of biological phenomena (see Chapter 5).

In practice, 'the system' to be comprehensively measured and modeled using integrated methods is usually a single cell. There are several reasons for this. Cells are the simplest of complex living systems, the smallest "units of life" (Boogerd et al. 2007). They are easy to individuate, with clear boundaries – membranes enclosing many molecular components. Their behavior is dynamic and context-sensitive, resisting simple reductive explanations in terms of a few genes or proteins. Most cells contain a nuclear genome, so focus on 'the cell' allows systems biology to build on and extend advances in genomics. Experiments in systems biology typically use single-celled organisms, such as bacteria and yeast, or cell lines, model organisms that resemble single cells (see Chapter 7). In its objects of study, systems biology reprises reductionist tendencies of mid-twentieth century molecular biology – and for similar reasons. Microorganisms are simpler and more tractable, and more data is available to construct models. 'The cell' is thus a major research focus for systems biology, and a central tenet of the field is that underlying networks control cell behavior. This "network hypothesis" is often stated in constitutive terms; e.g. "one can think of the cell as *made of* several superimposed networks" (Alon 2007, 97). More circumspectly, "it is possible to describe a cell through the set of interconnections between its component molecules" (de los Rios and Vendruscolo 2010, 4). Other systems biologists treat 'the cell' as merely the site of network action (critiqued in Bechtel 2010). But, however the cell is conceived, the basic assumption is that its behavior can be predicted and explained in terms of molecular networks that represent interactions among diverse molecules.

Systems biologists distinguish several types of network: metabolic, protein–protein, signaling, and transcription. These different network types differ in their components, timescale, or aspect of cell function targeted (Table 9.1). However, they are not conceived as isolated, but linked by shared components and a common site of operation: the cell (Figure 9.2). Metabolic networks consist of enzymes, substrates, and co-factors that

Table 9.1 The main types of cell network in systems biology. Timescales based on measurements from *Escherichia coli* bacteria

Network type	Key components	Timescale	Aspect of cell function	Inter-field connection
Metabolic	Proteins Small molecules	~1 msec	Energy generation and use	Thermodynamic principles
Protein–protein interaction	Proteins	~1 sec	Formation of functional complexes	Cell physiology
Signaling	Proteins Small molecules	~5 min	Interaction with external environment	Molecular biology of cells and tissues
Transcription	Proteins Small molecules DNA, RNA	~1 hr	Change in gene expression	Genomics

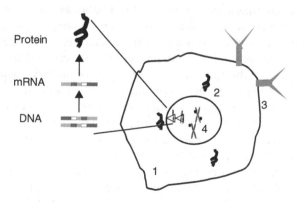

Figure 9.2 Location and principal components of the major types of cell network in systems biology today. (1) metabolic, (2) protein–protein interaction, (3) signaling, (4) transcription

participate in chemical reactions by which energy is transformed, stored, or released within a cell. Protein–protein interaction networks, as their name suggests, are comprised of proteins that directly bind one another, thereby bringing about effects on one or more components of the network. Signaling and transcription networks are closely related. The former mediate the relay of signals from a cell's external environment across the cell membrane, triggering a cascade of biochemical reactions inside the cell which culminates in some response to the signal. These cascades typically involve changes in gene expression. Transcription networks determine which genes are expressed, in the form of messenger RNA (mRNA) transcripts, in a cell at a given time. Also referred to as gene expression networks, transcription networks consist of diverse molecular elements: DNA sequences, RNA sequences, proteins, and small molecules (see Chapter 6). They are further classified, by timescale, into sensory and developmental networks. Sensory networks involve a short-term response to an environmental stimulus, and so overlap with signaling networks. Developmental networks control processes that operate throughout an organism's lifespan, though activity tends to be concentrated at early stages. So cell development is just one of several key phenomena of interest to systems biologists.[184]

Stem cell and systems biology share the aim of explaining cell development in terms of underlying networks of interacting molecules. In addition, both fields are interdisciplinary and emphasize the need for collaboration to achieve their goals.[185] Moreover, like stem cell biology, systems biology is in early stages of an ambitious project; its revolutionary potential is widely remarked, but has yet to deliver the goods. Aspirational commentary is prominent in textbooks and review articles. Typical examples include claims that "systems biology is the promise of biology on a larger and quantitatively rigorous scale, a marriage of molecular biology and physiology" (Szallasi et al 2010, ix), and that it is "set to revise many of the fundamental principles of biology, including the relations between genotypes and phenotypes" (Noble 2010, 1125). These hopeful remarks echo promises about stem cell research. Understandably, then, scientists in both fields are currently exploring ways they might join forces. In addition to reviews and commentaries, several workshops have recently been organized to foster collaborations.[186]

A key obstacle to this project is the methodological gap between mathematical modeling and stem cell experiments, visualized with great effectiveness in a 2010 conference poster (Figure 9.3). The goal is clear: underlying networks, mathematically modeled, are to explain and predict cell behavior. But the image of a living cell, grainy and irregularly-shaped,

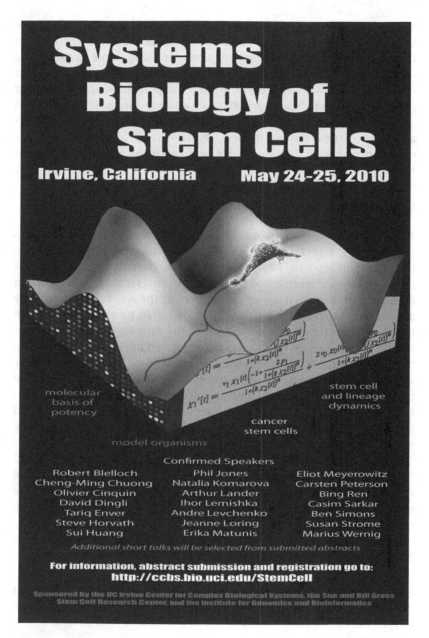

Figure 9.3 Visualizing collaboration: poster for a 2010 meeting aimed at integrating stem cell and systems biology (reprinted with permission of Arthur Lander, Center for Complex Biological Systems, Irvine CA)

sits awkwardly atop a landscape supported by systems concepts, equations, and high-throughput datasets. The iconography is telling; two fields are not yet integrated. But resources for doing so are in place. The next three sections propose a framework for collaboration between stem cell and systems biology. At the core of this proposal is an updated version of Waddington's landscape model, which was discussed in Chapter 6. This simple abstract model, I argue, mediates between stem cell experiments and mathematical systems models, revealing their interdependence for explanations of cell development. I begin with the 'underside:' cellular systems models.[187]

9.3 Mathematical modeling of cellular systems

A cellular systems model consists of a finite set of molecular elements $\{X_1, X_2... X_n\}$, representing DNA, RNA, proteins, and small molecules. Complexes of multiple components and functionally-distinct forms of one molecule are represented as distinct, so the set may be larger than a simple parts list. Each molecular element in the set is characterized by a value of a state variable $\{x_1, x_2... x_n\}$ at time t. In these models, the cell is defined as a complex system which at any time t is in a state $S(t)$ that is fully determined by the values of a set of variables $\{x_1, x_2... x_n\}$ representing the state of each molecular component.[188] The values of these variables exhibit numerous and diverse dependency relationships. Cell behavior, including development, is conceived as the result of changes in the values of state variables. A set of molecular elements, each described by a state variable, and dependency relations between the values of those variables, comprises a cell network model.

In this way, systems biologists aim to derive predictions about cell behavior, including developmental phenomena, from mathematical descriptions of interacting molecules. Insofar as these predictions are confirmed by experiments, the mathematical models that entail them can be said to explain the phenomena in question. The process of model construction begins with detailed description of a molecular mechanism. This mechanistic description is then simplified into a wiring diagram, which is next translated into a formal framework. Solutions within a formal framework correspond to vectors and attractors, and these vectors and attractors in turn define a 'landscape.' Figure 9.4 summarizes these steps, which are examined in detail in the next four subsections.

9.3.1 Molecular mechanism[189]

A molecular mechanism consists of diverse molecular components (proteins, DNA, RNA, and small molecules) that causally interact and are

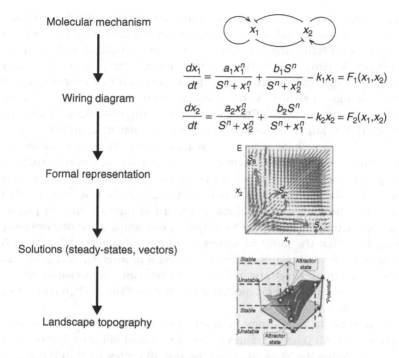

Molecular mechanism

Wiring diagram

Formal representation

Solutions (steady-states, vectors)

Landscape topography

$$\frac{dx_1}{dt} = \frac{a_1 x_1^n}{S^n + x_1^n} + \frac{b_1 S^n}{S^n + x_2^n} - k_1 x_1 = F_1(x_1, x_2)$$

$$\frac{dx_2}{dt} = \frac{a_2 x_2^n}{S^n + x_2^n} + \frac{b_2 S^n}{S^n + x_1^n} - k_2 x_2 = F_2(x_1, x_2)$$

Figure 9.4 Main steps of systems model construction. In the wiring diagram, inhibition is represented by an edge with a bar at the incoming end, activation by an edge with an arrow at the incoming end. Graphs on the right reprinted from Huang (2009) and Huang et al. (2009), with permission, respectively, from John Wiley and Sons, and Elsevier Press

spatio-temporally organized so as to constitute an overall working system, which performs some activity. Descriptions of molecular mechanisms narratively and diagrammatically exhibit the structure, spatio-temporal arrangement, and causal interactions of components. These descriptions explain by showing how some biological process works, in terms of its interacting parts. Mechanisms are discovered, and explanatory descriptions constructed, by means of experiments that manipulate parts of the concrete system under investigation. Cellular systems models presuppose a mechanistic description of the cell behavior of interest, based on such experiments.

9.3.2 Wiring diagram

The first step of model-construction is to simplify the mechanistic description into a wiring diagram. Choice of what to include is guided by selection of a state variable. The state variable is usually concentration or some proxy thereof, such as gene activation or expression level. The wiring diagram represents all the molecular elements and their interactions relevant to determining the value of the state variable of each element. In network terms, the wiring diagram's nodes correspond to molecular elements, edges to interactions. Interactions relevant to determining the value of a given element's state variable are incoming edges and those influenced by the value of that variable are outgoing edges (Figure 9.4, upper right). Much of the detail of the molecular mechanism is omitted, including cellular structures, spatial organization, and molecular structure of the elements themselves. All that remains is the structure of interactions relevant to determining the value of the state variable of each molecular element. In deterministic models, it is assumed that values of the state variable of each element at time t suffice to predict values at later times. Another assumption, not restricted to deterministic models, is that the range of values taken by the state variable of each molecular element relates to the cell behavior of interest, i.e. the overall phenomenon for which the molecular mechanism is responsible. Absent any such relation, systems models could not explain cell behavior, even in principle.

State variable values may be qualitative or quantitative, discrete or continuous. All possible values of the state variable for each molecular element define the 'state space' of the system represented in the wiring diagram. Each element corresponds to one dimension of the state space, and each point within that space corresponds to a different combination of values for the molecular elements in the network. At any given time the system as a whole occupies some point in state space: $S(t) = [x_1, x_2 \ldots x_n]$. The wiring diagram in Figure 9.4 depicts a very simple system of just two elements, X_1 and X_2, each of which self-activates while inhibiting the other. The state space of this simple network has two dimensions. Because of its simplicity, it is easy to see that this wiring diagram represents a system with two possible qualitative outcomes: high concentration of X_1 and low concentration of X_2, or the converse. This simple structure is a common motif in cell developmental networks. In a transcription network, the state variable is a measure of gene expression at the level of mRNA transcripts. Molecular elements X_1 and X_2 represent genes expressed as mRNA, and their "wiring" connections represent the effects, direct or indirect, of gene expression.

9.3.3   Formal framework

Next, the wiring diagram is translated into a formal framework. A number of different frameworks are available, including ordinary differential equations (ODEs), directed graphs, Boolean networks, and Bayesian networks.[190] Directed graphs involve minimal transformation; graph theory is simply applied to the wiring diagram itself. However, directed graphs are insufficient to model network dynamics, so this formalism cannot be used to predict or explain cell development. Bayesian networks are a powerful tool for inferring probabilistic dependency relations among values of variables. Though essentially statistical, under certain conditions Bayesian formalism can be used to predict network dynamics. However, these conditions include a great deal of information about the system under investigation, which is rarely available for molecular mechanisms of cell development. So models of cell development, at least for now, tend to eschew Bayesian formalism. The Boolean framework is much more versatile, representing gene expression as a binary ON/OFF (1, 0) and interactions among elements as rules of propositional logic incorporating connectives AND, NOT, OR. Because values of variables are discrete in the Boolean framework, however, this formalism does not lend itself to a landscape representation. For my purposes, then, it can be set aside.

ODEs are the most commonly used formal framework for cellular systems models, both in systems biology generally and for stem cell phenomena in particular.[191] A system of ODEs is a continuous, deterministic model that describes how the state variable of each molecular element of the wiring diagram changes over time. The general form of an ODE model is:

$$\frac{dx}{dt} = f_i(x_1...x_n), \; i = 1,...n$$

where f_i denotes a rate law specifying how the value of the ith element's state variable depends on the values of variables of other elements in the network. If the state variable is concentration or molecule number, then the form, parameters, and constants of f_i catalog the processes by which the ith element of the network is produced and degraded. A system of ODEs represents the change in concentration of each molecular element over time.

In the two-element system shown in Figure 9.4, the minimal systems equations are:

$$\frac{dx_1}{dt} = \frac{a_1 x_1^n}{S^n + x_1^n} + \frac{b_1 S^n}{S^n + x_2^n} - k_1 x_1$$

$$\frac{dx_2}{dt} = \frac{a_2 x_2^n}{S^n + x_2^n} + \frac{b_2 S^n}{S^n + x_1^n} - k_2 x_2$$

where the first term on the right-hand side represents activation of gene expression, the second represents inhibition of gene expression, and the third represents degradation of mRNA transcripts. The a, b, and k parameters represent the rate, or strength, of these different kinds of interaction, while S and n parameters describe the form of regulatory interactions, fixed by the biochemical stoichiometry of the system, but usually unknown.[192] Stochastic models are even more fine-grained, describing the probabilities of individual reaction events that influence the number of molecules of elements in the network. These models include systems of ODEs as a special case. Much of what follows therefore applies to cellular systems models of either sort.

9.3.4 Solutions and derivation

Solutions to a system of ODEs define features of a 'landscape' in state space. One kind of solution that can often be found is a "steady-state" at which there is no change in values of variables, that is the left-hand side of each dynamic equation is set to zero. A given system of ODEs may have one, several, or no steady-states.[193] The system depicted in Figure 9.4 has three, shown as S_A, S_B, and S_C. Another kind of solution that can often be calculated or approximated is a "local solution" given a set of initial conditions; values for the state variable of each molecular species at a given time t. Each local solution describes exactly how the system, as modeled, will change over the next small increment of time. A set of initial conditions corresponds to a point in state space. Local solutions correspond to tiny vectors in state space originating at that point. A systematic canvassing of local solutions creates a pattern of tiny paths in state space, "like the hair on the head of a new military recruit" (Conrad and Tyson 2010, 109).

The orientation of vectors in state space determines whether steady-states of the system are stable or unstable. A stable steady-state is defined as one toward which vectors converge, an unstable steady-state as one from which vectors radiate away. Two of the steady-states shown in Figure 9.4 are locally stable (S_A, S_B); the third is unstable. This arrangement of steady-states conforms to the intuitive result, that one or the other of X_1 and X_2 is expressed, but not both. This simple network structure is therefore termed a "bistable genetic switch." A landscape is produced by adding the dimension of 'stability' to the state space. For a two-dimensional state space, this is visualized as 'elevation' of the surface representing the state of the system (Figure 9.4, bottom right). Systems of higher dimensionality are not so easily visualized, but the principle remains the same. Landscape topography represents the relative stability of each possible state of the system as a whole. The arrangement of vectors and steady-states in state space determines landscape topography. In this sense, cellular systems models derive the landscape 'from the bottom-up.'

9.4 Waddington's landscape revisited

This modeling approach is, in effect, a principled elaboration of the underside of Waddington's landscape (1957). The latter model, to briefly recap, is a two-dimensional diagram representing pathways of organismal development as an inclined surface etched with branching valleys (see Figure 6.3). The axis projecting outward represents time. The horizontal axis represents phenotype ordered by similarity (the measure of which is left unspecified). The vertical axis represents order of development, correlated with time such that the developmental surface is tilted. In this model, development is a force that operates like gravity on organisms and their parts. But its operation is structured into branching tracks, with a single starting point leading to multiple end-states. So the landscape, as a whole, represents developmental options progressively restricted over time.

9.4.1 Bottom-up regulatory networks

Waddington hypothesized that developmental pathways are determined by networks of interacting genes (see Figure 6.4). His speculative account of the role of genes in development is markedly similar to the prevailing view in systems biology today. In particular, the concept of feedback was crucial. Waddington's unified view of genetics and development, briefly, was as follows. Chromosomes provide a context for gene–gene interactions in the generation of physiologically active products, which interact with one another in biochemical pathways. Biochemical pathways, in turn, feed back to influence chromosome structure and gene–gene interaction. Though coarse-grained in comparison to recent models of transcriptional regulatory networks, this interactive account of the role of genes in development is prevalent today. Cellular systems models replace the genetic "pegs" that anchor Waddington's landscape with a dynamic network of interacting molecules.[194]

Although the 'underpinning' networks are similar, cellular systems models contrast with Waddington's landscape in three significant ways. First, they represent cells rather than organisms or organismal parts. Though Waddington himself did not interpret the landscape model as representing cell development, there is no obstacle to doing so. In the original model, the developing entity is left unspecified. Today, after decades of progress in molecular and cell biology, the ball poised at the top of the landscape is naturally interpreted as a cell. The second contrast cuts deeper. Cellular systems models entail predictions, while Waddington's landscape does not. For Waddington, the unified, interactive view of development and genetics was a speculative guide to inquiry, not an established theory or confirmed hypothesis. The landscape model was intended for "a context in which it is more important to employ a system of thought which is flexible and of wide application than to search for a precise formulation of a narrower viewpoint" (1957, 31). To play its unificatory role, the model was constrained to be simple, rather than an accurate or principled representation of animal development.

Cellular systems models, in contrast, are designed to be predictive. The formal framework of ODEs entails myriad predictions about how the state of the system will change, given a set of initial conditions. This allows one to derive many predictions about how the system as a whole will respond to manipulations that change the state variable of one or more molecular elements. Applied to cell development more generally, systems models predict that mature cell states correspond to stable attractors, and branch-points to bistable local attractors. In practice, however, predictions of cellular systems models usually deviate from experimental results. They are currently used primarily to detect as-yet-uncharacterized components of molecular mechanisms and to suggest hypotheses for concrete experiments.

The third contrast concerns explanatory significance. As with prediction, Waddington's model was not intended to be explanatory, but to provide a framework for articulating explanatory concepts. Cellular systems models, however, offer a 'bottom-up' derivation of the landscape itself, from an underlying molecular network. Note that it is the network as a whole that determines landscape topography, not any "master molecule" within it. Stability of cell states is determined by an entire regulatory network. A regulatory network consists of diverse components (including DNA, RNA, protein complexes, and signaling molecules) that together produce a pattern of gene expression. Each pattern of gene expression exhibits a certain degree of stability in a given environment. Cell states can therefore be characterized in terms of stability and robustness to environmental change. Robustly-stable cell states are construed as equilibria of the system, actively maintained by interactions among its diverse components. So stability and robustness are emergent properties determined by the state of the whole cell. Of course, genes are included in the regulatory network (see Chapter 6). But they are not the "ultimate ground" that determines features of the epigenetic landscape.

If explanation tracks mathematical derivation, then cellular systems models can, in principle, explain cell behavior, at least in response to minor perturbations. Yet, in practice, systems biologists rarely discuss their models and results in terms of explanation. One reason for this is that systems models of cell development are works-in-progress, and therefore tend to be used as hypothesis-generating devices rather than as grounds for explanation. There is more to the explanatory situation, however, than incomplete datasets. The derived stability landscape suggests that the trajectory of cell development is controlled by the regulatory system as a whole. Structural features of development, the inclined surface with branching valleys descending, punctuated by 'mountain lakes' representing intermediate stages, should therefore emerge from the bottom-up. But such a derived landscape can only explain cell development insofar as it corresponds to the results of experiments that characterize those pathways. Here the perspective shifts to the top-side of the landscape surface, the terrain of stem cell experiments.

9.4.2 Top-side experiments

As discussed in Chapter 6, Waddington's landscape has recently been co-opted by stem cell biologists for three purposes: to visualize shared background assumptions, express generalizations about experiments, and correlate cell state with developmental potential (see §6.6). Each of these roles bears on explanations of cell development. The most central shared background assumption concerns the stem cell concept. If the developing entity is interpreted as a cell, the landscape model visualizes key features of this concept: a single undifferentiated starting point, with the potential to develop along a variety of pathways, gradually restricted as development proceeds, ending with stable, mature cell types. From this perspective, then, the landscape model represents the stem cell concept. It also provides a background context for representing stem cell experiments. At their most basic level, stem cell experiments consist of removing cells from their physiological context and placing them in new environments. This is essentially the strategy of experimental embryology, applied to cells rather than embryos or tissue explants. Waddington's landscape is thus an appropriate framework for representing stem cell experiments.

Modified landscape diagrams (such as those depicted in Figures 6.3 and 6.4) represent generalizations about reprogramming experiments, with details of particular experiments (organismal source of cells, ingredients of culture medium, specific genes or proteins added, method of delivery to the nucleus, and so on) omitted. Each arrow on the landscape represents many specific experiments, which use different cells, different techniques for maintaining and manipulating them, and different ways of measuring outcomes. Different classes of reprogramming experiment are distinguished not by procedural details, but by the tracks they describe on the landscape, each corresponding to a different deviation from normal development. In this way, the landscape model compresses thousands of reprogramming experiments into a few generalizations, visualized on the landscape. These generalizations suggest possible explanations for experimental outcomes. Stem cell biologists look to underlying molecular mechanisms to explain the experimental generalizations represented on the landscape. Reprogramming experiments also reveal components of these mechanisms by revealing differences made to cell development by specific DNA, RNA, and protein molecules, as well as chemical modifications of DNA and features of the chemical environment. A third use of the landscape model is to represent the correlation of molecular features of cell state and developmental potential. These correlations set the stage for MEx of cell development.

The landscape model also subtly biases such explanations, by insulating certain assumptions from critical scrutiny. This is neither unusual nor necessarily pernicious. Like all models, Waddington's landscape encourages some insights while obscuring others. But it is important to recognize such biases. One example concerns the unidirectionality of development in time, represented in the landscape model as the incline. Change in cell state with time corresponds to motion along the developmental surface. In the intuitive visualization the model is designed to elicit, it is easy to suppose that ordinary gravitational forces obtain. The direction of motion is determined as if cell and inclined surface comprised a simple mechanical system. The principles directing development are tacitly analogized to laws of mechanics. Yet the latter's exactness and predictive power nowhere enter into the model. The assumption that cells develop over time as ineluctably as balls rolling down an inclined plane figures in the model as a law of development – but not one that can be tested by deriving predictions from the model. This encourages the assumption that directionality of development (progress down the slope) needs no explanation.

The landscape model also excludes hypotheses of extreme adult stem cell plasticity. 'Adult stem cell plasticity' refers to the idea that (some) stem cells found in adult tissues have developmental potential extending beyond those tissues. For reasons discussed in Part I, hypotheses about cells' developmental potential resist decisive test, so it is difficult to exclude them based on experimental evidence (see Chapter 4). However, *modest* plasticity hypotheses are well-supported. For example, liver stem cells can give rise to cells of other organs, given an appropriate environment. *Extreme* plasticity hypotheses go further, implying that developmental pathways are highly malleable: blood can make brain, liver can make heart muscle, bone can make kidney, and so on.[195] The current consensus is that adult stem cell plasticity is not extensive. The landscape model, fixed and sculpted by deeply-entrenched valleys, expresses this consensus diagrammatically: hypotheses of extreme, widespread adult stem cell plasticity cannot be visualized on such a landscape. Overall, Waddington's model constrains acceptable hypotheses about cell development to accord with the basic principles it embodies: development is a continuous, one-way process; cell potency is restricted over time by a robust and orderly pattern of binary 'choices;' terminally-differentiated states are stable. It thus provides a convenient framework for representing the correlation between developmental potential, represented by position within the branching pathways descending the slope, and molecular characteristics that define cell state.

9.5 Explaining cell development

The two updated versions of Waddington's landscape, from cellular systems models and representations of reprogramming experiments, can now be brought together. On a 'derivational' account of scientific explanation, cell development is explained by deriving the landscape's incline and contours from regulatory 'circuitry,' and showing that experimental generalizations match the predicted trajectories. On this view, the two sides of Waddington's landscape map onto each approach's explanatory role. Cellular systems models express hypotheses about underlying molecular networks, while generalizations representing stem cell experiments articulate phenomena to be explained. Predictions from the former are tested against data from the latter and, insofar as they match, cell developmental phenomena are explained. Both are necessary, but the mathematical model is the explanans, the locus of explanatory power. The role of concrete experiments is to articulate the explanandum-phenomenon, preparing the ground for an explanation-by-derivation.

This explanatory picture, despite its intuitive appeal, fails to capture several important features of the case at issue. This is not to say that the derivational account of explanation is inadequate across the board. But for cell development, and the two approaches under consideration, the traditional view gets things backward. Mathematical models of cell development depend on concrete experiments in two ways: for construction, and for use in making predictions about cell development. The first is obvious: mathematical models assume a description of a molecular mechanism (§9.3.1). These descriptions are based on results of concrete experiments, which identify the molecular elements of a network and characterize causal interactions among them. Rate laws in a system of ODEs approximate the latter, often with idealizations to improve computational tractability. But none of the molecular dependency relations implicated in mechanisms of cell development is derived from first principles of (say) thermodynamics.[196] The details of production, activation, inhibition, and degradation represented in the rate equations for gene expression are based exclusively on data from concrete experiments. Moreover, these equations do not reliably generalize beyond the cellular context in which data are generated. This is because the same molecular components often interact differently in different contexts. In particular, the same molecules may interact differently in developing than in mature cells, in different species, or different mature cell types. So molecular networks operating in developing cells require data from the cells in question, in order for equations to be formulated. This restriction does not hold for cellular systems models that play a more speculative role, proposing hypotheses for experiments to test. But in their explanatory capacity, cellular systems models are restricted to specific molecular mechanisms. The equations comprising these mathematical models, at best, regiment the results of concrete experiments.

Second, systems models depend on concrete experiments to make predictions about cell development. This is less obvious than their dependence for model-construction. But, once constructed, a cellular systems model entails only predictions about combinations of values of the state variable, given a set of initial conditions (see Figure 9.4). These predictions are about cell states, not cell development. The relative stability of different states, reflected in the arrangement of vectors in state space, does not determine developmental pathways. In order to make predictions about the latter, we need a way of orienting ourselves on the developmental landscape, so to speak. A developmental order must be superimposed on the landscape of vectors and attractors. That is, predictions about development require that we first identify the formal features of Waddington's model: undifferentiated start-point, disparate endpoints, and intermediate stages. Most important is the starting point of a developmental trajectory, which provides a point from which to orientate on the developmental landscape. This is just to say that predictive models need to identify the position of a stem cell in the overall landscape. Concrete stem cell experiments, which correlate molecular state and developmental potential, do exactly this. These correlations are the result of many stem cell experiments, conveniently summarized on the landscape. Supplemented with such experimental generalizations, which connect cell state and developmental potential, many predictions about cell development follow from the system of equations from which landscape topography is derived. In this way, mathematical systems models depend on concrete experiments to make predictions about cell development.

If the derivational account of explanation is correct, then the role of concrete experiments is not limited to articulating the explanandum. Mathematical systems models are constructed using experimental data and depend on other experimental results, linking cell development to molecular states, to derive predictions about cell development. But there are further difficulties for the derivational account in this case. Explanations of cell development describe molecular mechanisms, the details of which are worked out by concrete experiments that manipulate parts of cellular systems. So the descriptions presupposed by cellular systems models *are themselves explanatory*. The mode of explanation involved here is not derivational, however, but mechanistic. MEx describe mechanisms, complex systems composed of jointly interacting parts, that together constitute a phenomenon of interest – in this case, cell development. Chapter 5 argues that MEx is a species of part-whole explanation: a higher-level phenomenon is explained by description of its lower-level parts and their interactions. Such a description explains *how* the overall mechanism works. What makes a mechanistic description *explanatory* is unification of levels: exhibition of how working parts together constitute the overall mechanism.

MEx are therefore not derivations, but unify causal processes at different levels of organization – in this case cells and molecules. A successful MEx exhibits how working parts constitute an overall mechanism, so we can see (often literally, via a diagram) how two levels map onto one another. In this case, the levels are cellular phenomena and underlying molecular networks. It is not enough to specify how changing one or a few molecular components changes the whole system, which an experiment can reveal on its own. A MEx must 'put it all together,' exhibiting how all the inter-related parts together give rise to the overall working system. Typically, the display is in the form of a diagram that shows key aspects of all the relevant parts, including how they mesh and the properties that allow them to do so.

This joint account of MEx clarifies the role of mathematical models in explanations of cell development. For some high-level systems, working parts 'sum together' or combine in a way that is easy to specify, and there is no need for mathematical modeling. Many explanations in molecular, cell, and developmental biology are of this sort. But cell development involves very complex dependency relations and multiple feedback loops. Revealed by piecemeal experimentation on concrete systems, molecular interactions underlying cell development flummox our ordinary intuitions as to how the two levels are related. Even a three-element "core network" of pluripotency genes (see Chapter 6) operates in a manner that is hard to parse intuitively. The difficulty is compounded by the need to understand how transcription factors 'hook into' the gene expression system of a cell so as to alter its morphology, differentiation potential, and other traits. Because of this complexity, we need technical help to understand the relation between working parts and overall system, and therefore to construct successful MEx of reprogramming and other cell developmental processes. Mathematical models specify the relation between interacting genes, proteins, RNA, and small molecules on the one hand, and overall system behavior on the other.

Though crucial, this explanatory role for mathematical models is quite circumscribed. The specific details of a MEx, including molecular details, causal interactions, and correlation of molecular traits with cell developmental potential, are grounded on concrete experiments. Mathematical models are limited to making the interlevel connection explicit. But MEx of cell development cannot succeed without this step. So both concrete experiments and mathematical models are needed to mechanistically explain cell development. In this sense, systems biology is necessary for progress in stem cell biology. But MEx of cell development are grounded primarily on concrete experiments, not on mathematical models.

9.6   Models and experiments

The landscape model thus mediates between stem cell and systems approaches to understanding cell development. The previous section shows their interdependence in MEx of these phenomena. Mathematical systems models depend on concrete experiments, both for construction and for prediction. Though experiments on concrete biological systems are the source of mechanistic details about cells, molecules, and their environments, these explanations also require mathematical models to explicate the link between interacting molecules and behavior of the cell as a whole. This section discusses some implications of this result for philosophical accounts of models, theories, and experiments.

9.6.1 Unifying models

The landscape case demonstrates that simple abstract models can play a significant role in science apart from prediction, explanation, and hypothesis-generation. On the systems approach, the stability landscape is a derivational consequence of very simple mathematical models. As used by stem cell biologists, the landscape model abstracts away the details of experiments aimed at elucidating specific mechanisms. In neither case is the landscape model used to predict or explain cell development. Instead, it plays a unifying role – as Waddington intended, albeit for different fields. In stem cell biology the model depicts a shared starting point for diverse experiments, relating them all to the same basic principles of development, and thereby facilitating construction of MEx. Similar models may be found in other experimental fields aimed at constructing MEx of complex systems, such as immunology, psychology, and neurobiology. At the emerging interface of stem cell and systems biology, the landscape coordinates experimental and modeling approaches, indicating their complementary roles in explanations of cell development. This simple, abstract model facilitates formation of a new interdisciplinary community.

9.6.2 Microarray experiments

The above shows that concrete experiments ground the 'bottom-up' strand of systems biology insofar at it aims to explain cell development. But the 'top-down' approach is also associated with experiments – specifically high-throughput technologies that comprehensively measure molecular constituents of a cell (§9.1). Commentaries on systems biology often give the impression that such "data-driven" experiments are the primary, or even sole, complement to mathematical modeling, with more traditional experiments of molecular, cellular, and developmental biology relegated to a minor role (e.g. Aggarwal and Lee 2003, Krohs and Callebaut 2007). Such claims emphasize systems biology's novelty, breaking with established lines of biology research. However, a closer look at top-down systems experiments reveals dependence on, and continuity with, mainstream experimental biology.

High-throughput omics technologies are diverse, and continually proliferating.[197] Here I focus on the earliest and most influential, DNA microarrays, as a representative example. DNA microarrays are constructed by attaching thousands of short DNA sequences in a gridlike pattern to a glass slide or "chip." The source of these DNA sequences is typically a 'library' of genes from the species or strain of interest, which are commercially available for well-studied organisms. A microarray experiment uses a chip to detect genes expressed in a cell under particular conditions, which vary from experiment to experiment. mRNA in the target cell (the result of transcription) is extracted and used as a template to make complementary DNA sequences, which are then labeled with fluorescent dye. Most microarray experiments involve comparison of cells under different conditions; in these cases, each RNA-to-DNA treatment is labeled with a different dye (one red, one green), and the two are mixed together. Labeled DNA is then incubated with the chip, allowing complementary sequences to bind one another. After washing to remove unbound DNA sequences, the chip is scanned with a laser to detect each kind of fluorescence. Fluorescent red or green spots on the chip indicate genes expressed in the target cell(s).

Microarray data require extensive processing by automated instrumentation, from photomultipliers to statistical software programs. Interpretation of the resulting matrices, representing patterns of gene expression in a color-coded grid, requires further analysis still. So bioinformatics and computational biology accompany these technologies. Yet, at the core of a microarray experiment are the familiar operations of molecular biology: nucleotide coding, specific binding of complementary molecules, and biochemical modification.[198] The main difference is that expression of many, rather than one or a few, genes is measured, yielding a 'snapshot' of the overall cell state.

The notion of experimenting communities (Chapter 8) is helpful in clarifying the relation between high-throughput experiments to mainstream molecular biology. Though their basic principles and core design are very similar, as modes of participation in a wider experimenting community the two differ considerably. The usual practice in an experimental biology laboratory is to focus on one, or a few, molecular constituents of a mechanism of interest, tracing the details of meshing interactions with other molecules, as well as effects on the overall mechanism. Each laboratory group investigating a particular mechanism contributes a tiny fragment of a more complex picture, 'biased' so as to showcase one or a few molecules. Larger explanatory models are assembled by gradual integration of these fragments, aided by unifying frameworks like Waddington's landscape. The process is decidedly non-systematic, sensitive to the contingencies of personal interactions at conferences, local expertise, and exchanges of laboratory members. High-throughput methods 'compress' experiments, which on the traditional approach would be distributed across many laboratories, to a single benchtop. As the number of molecules measured far exceeds the number of laboratories investigating a given mechanism, omics technologies also provide more comprehensive datasets than an entire experimenting community working in the traditional way. It is in this sense that systems experiments are "data-rich" and "unbiased," as their proponents claim.

But high-throughput methods are not a royal road to comprehensive MEx. Robustness and comparability of results across laboratories must still be worked out, just as in traditional biology experiments. Omics datasets are relative to the conditions of cells measured, which vary in myriad ways across (and even within) laboratories. So community-level practices that integrate results across laboratories are still required. Moreover, regulatory models based on data from high-throughput experiments do not, on their own, yield MEx of cellular phenomena. A variety of top-down modeling strategies can be used to infer regulatory network structures from omics datasets. Like their bottom-up counterparts, however, these models must be coordinated with mechanistic details revealed by molecular and cell biology experiments.[199] The top-down perspective also shows systems and mainstream experimental biology to be interdependent.

Other than commitment to mathematical modeling, systems biology's most radical break with mainstream experimental biology is in social organization of experiments. In other respects, its core methods, concepts, and explanatory standards are continuous with molecular and cellular biology. Accordingly, omics technologies are eagerly taken up by many stem cell researchers.[200] These high-throughput methods are instruments of an expanded, experimenting community of molecular and cellular biologists, which invites new contributions from mathematics, informatics, computer science, and engineering.

9.6.3 General principles

Some systems biologists, namely those committed to systems-theoretic principles, are not satisfied with this piecemeal situation (§9.1). Alongside omics technologies, they advocate new theoretical foundations for experimental biology: quantitative design principles for living organisms. For example, Kauffman (1969, 1993) seeks mathematical laws, analogous to statistical mechanics, to confer stability and order on the myriad molecular interactions of metabolism and development. Kitano (2002) speculates that comparisons of network structures and emergent biological outcomes could yield a general system for understanding life; something like a periodic table or evolutionary tree of functional design patterns. Such accounts would revolutionize biology, demanding changes in core concepts and explanations. A number of systems-theoretic biologists have proposed stem cell biology as a field ripe for revision. Arguing that "the stem cell concept... is holding us back," they propose successor accounts in terms of landscape attractors or chaotic systems (Lander 2009; also Huang 2009, Kaneko 2011). An influential example, especially among philosophers, is Stuart Kauffman's theory of orderly development emerging from complex interacting networks.

Kauffman and colleagues seek principles to account for orderly features of development: stability of phenotypes and pathways, as well as robustness of the developmental process itself, unfolding across generations and phyla. Their approach is not to manipulate biological entities themselves, but to study dynamic properties of formal Boolean networks with analytical tools and computer simulations. Kauffman's networks consist of nodes (termed "genes") assigned values of 0 or 1 (off or on), governed by rules formulated in terms of Boolean operators (and, or, not) which map inputs from other nodes at time t onto outputs, node values for time t+1. By implementing the rules over long intervals in many replicates, from many different initial states, dynamic properties of a network emerge. Using this approach, Kauffman explores the effects of variation in network size, connectivity (K, or number of inputs per node), and prevalence of "noise" (randomly-introduced changes to a network). Simulations show that large networks in which each node has few inputs (K = 2) exhibit orderly behavior that mimics cell development. In particular, node values quickly converge on a cyclic pattern of repeating states that is robust to small perturbations: an attractor. By systematically varying initial states, different attractors for a given network are revealed. Importantly, network states are not all equidistant. The options for direct transition from one state to another are constrained by the underlying structure of interacting nodes.

The parallels between Kauffman's simulations and organismal development are suggestive. In particular, networks of 10^5 components with connectivity K = 2 exhibit 317 stable states – roughly the number of cell types in an adult human. Given representational assumptions that identify nodes with genes, node values with all-or-nothing gene expression, interactions with the effects of molecular mechanisms of gene expression, network states with cell states, and cell types with attractors, Kauffman's models predict a number of features of cell development: initiation, robustness to noise, multiple cell types, and branching pattern. On these grounds, Kauffman claims that network structure provides "a theory of differentiation" that explains the "origin, sequence, branching and cessation" of cell development in terms of attractors (1969, 438). The similarity to Waddington's landscape is obvious.[201] But Kauffman's account is far more sweeping than derivation of a stability landscape from details of a molecular mechanism. His models omit all molecular details; interactions linking nodes, as well as Boolean rules, are randomly assigned. The order that emerges from network structures is independent of any molecular mechanism; indeed, of any biological entity whatsoever. On Kauffman's account, cell development is explained by general principles of complex, self-organizing systems. Core phenomena of development, including stem cell capacities, are recast in terms of dynamical systems theory: attractors, noise, and robustness. The result is a new concept of stemness, replacing cells with systems, which characterizes the "essence of multipotency" as an attractor in a multistable landscape (Huang et al. 2007, 695).

On its own terms, this systems-theoretic account of cell development exemplifies a legitimate and promising research program. Systems biologists, often in collaboration with the closely-allied field of synthetic biology, aim to articulate general design principles that systematically characterize the domain of possible network structures underlying biological phenomena. The resulting 'how-possibly' explanations (still very much works-in-progress) are neither causal nor MEx, but more like covering-law explanations. The main explanatory targets of these early-stage 'design explanations' are multiple forms of robustness: adaptation to environmental change, insensitivity to kinetic parameter change, and graceful degradation rather than catastrophic malfunction. Network structures achieve robustness in one or more of these senses via feedback, redundancy, modularity (in the sense of relative isolation of subsystems), and/or stochastic heterogeneity (noise). Explanatory power is localized to features of interactions among components, independent of specific parameters governing actual reactions. In a few cases, systems-theoretic models have been shown to fit well with experimental results, for example a "switch" between two blood cell lineages (Enver et al. 2009, Huang et al. 2007). However, these theoretical ideas have as yet made little impact on stem cell biology.

Stem cell biologists' indifference to systems-theoretic accounts of development is particularly striking when contrasted with their eager uptake of omics technologies. In Kauffman's case, one factor is his assumption that genes control development, which clashes with the aims and explanatory norms of stem cell biology (see §6.7).[202] But 'genomic reductionism' is not the only issue. As argued earlier, stem cell biologists need conceptual tools to coordinate the results of diverse experiments and construct robust MEx. The landscape model is one such tool, which some stem cell biologists have adopted. Why not go further, with unifying general principles providing a framework for more specific MEx? There are, I think, two reasons. The first is that systems-theoretic accounts do not, so far, offer any advance on experiments. Models such as Kauffman's at best conform approximately to experimental results; they yield no exact or surprising predictions. So, for experimental biologists, systems-theoretic reformulations of stem cell concepts merely redescribe familiar results, rather than prevailing in severe tests. Moreover, the experimental results to which models' predictions are compared are usually far from the cutting-edge of stem cell biology, applying familiar techniques to tractable cell populations. Even a much greater degree of fit in results would not impress most stem cell researchers, without some link to the 'growing edge' of the field. But, unlike systems models that test our understanding of particular mechanisms, systems-theoretic redescriptions of stem cells, by their very generality, offer no specific guidance for further experiments. So these accounts do not move ongoing experimental inquiries forward. Though coherent on their own terms, they offer no attractions for experimentalists.

The second reason is deeper. Why build and study an abstract model if the concrete objects of interest – stem cell lines and organismal parts – are accessible? One reason is that abstract models can reveal possibilities not realized in the actual systems. Systems-theoretic accounts aim to articulate principles for understanding this wider possibility space. In many fields, this kind of systematic understanding is the main goal of inquiry. But for stem cell biology, the overall goal is therapeutic: to use stem cells to treat a wide range of pathologies, such as heart attack, blindness, cancer, neurodegenerative disease, paralysis, diabetes, complications arising from organ transplantation, and many more. Chapter 2 introduced the idea that stem cell biology does not need unifying theories. The intervening chapters have further articulated the epistemic aims and standards at work in the field, with emphasis on experimental methods, mechanistic explanations, and diverse kinds of models. These partial and piecemeal studies proceed in the absence of a unifying theory. Instead, the unifying role is played by clinical aims, which spur collaboration and integration of diverse models and methods. Therapeutic values are, in this sense, more fundamental for stem cell biology than theories. This idea is explored further in the final chapter.

9.7   Conclusion

Stem cell and systems biology are distinct fields, with much to gain from one another. But their methods are very different. Contrasting methods, together with entrenched assumptions about the respective roles of abstract theory and concrete experiment, could impede cooperation between them. The relation between mathematical modeling and concrete experiment is often cast in oppositional or competitive terms, with one or the other approach predominating. A related idea is that experimentation prevails at early stages, with theory asserting dominance as a science matures. But the situation in biology is more nuanced. Waddington's landscape helps us visualize the interplay of stem cell and systems biology, offering a metaphorical surface on which the two approaches can meet as partners. This simple abstract model illuminates the roles of mathematical modeling and concrete experimentation in constructing explanations of cell development. Omics technologies are largely continuous with molecular and cellular biology, while more thoroughgoing systems-theoretic treatments have little impact on stem cell research. Abstract modeling, though crucial for explanations of cell development, is no substitute for experiment.

10
Clinical Values

10.1 Introduction

This chapter examines the relation between stem cell biology and clinical medicine. Unlike the emerging interface with systems biology, the association of stem cell research and clinical medicine is quite entrenched. Indeed, there is a sense in which clinical values are constitutive of stem cell biology. Clarifying this statement draws together many results from earlier chapters. So this final chapter provides a capstone to this book, as well as indicating areas for future work.

The role of clinical values in stem cell biology is relevant to long-running debates over scientific objectivity. A constitutive role for clinical values in stem cell biology conflicts with the traditional ideal of "value-free science:" pure knowledge without the biasing influence of ethical, political, and social values. The value-free ideal has been sharply criticized, notably by proponents of socially-relevant philosophy of science. Its critics, though variously motivated, share a concern that philosophy of science should produce results relevant for policy. The following sections extend these criticisms to the stem cell case, and discuss their implications for the future of stem cell research.

10.2 Translational research

Today, the relation between stem cell research and clinical medicine is conceived primarily in terms of "translational research" or "clinical translation," ideas that are supplanting the traditional dichotomy of pure-vs-applied science. Instead of two ordered activities (pure science followed by its application), the translational view emphasizes continuous mutual support between laboratory and clinic: "Basic scientists

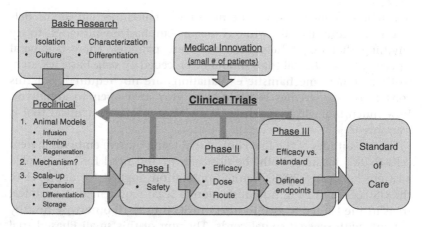

Figure 10.1   Clinical translation schema. Reprinted from Doulatov et al. (2012), with permission from Elsevier Press

provide clinicians with new tools for use in patients and for assessment of their impact, and clinical researchers make novel observations about the nature and progression of disease that often stimulate basic investigations" (NIH 2008). Often represented in 'flowchart' form, the translation process exhibits feedback relations, mirroring the biological processes that are objects of this research (Figure 10.1). But unlike their biological counterparts, such 'social mechanisms' are understood only vaguely, without experimental grounds. The crisp boundaries of flowchart stages do not correspond to real situations, which are murky and contentious. For example, even within a particular laboratory, the boundary between "basic" and "preclinical" research is difficult to specify. Flowcharts and roadmaps for translational research are idealizations, which policy-makers, researchers, and clinicians must work to realize and connect with empirical results.

In stem cell research, principled guidelines for clinical translation are greatly needed, as the field presents "a potentially lethal cocktail of desperate patients, enthusiastic scientists, ambitious clinicians and commercial pressures" (Watt and Driskell 2010, 159). High stakes and diversity of interests in play, compounded by cultural battles over human reproduction and healthcare, make expectations of costs and benefits difficult to assess. One key debate concerns when clinical trials should begin. This is a general problem for translational research, as clinicians and laboratory scientists often arrive at conflicting judgments owing to their different practices and professional responsibilities. Laboratory researchers frequently stress the need for detailed mechanistic models to warrant clinical trials, while clinicians emphasize willingness to try anything that might help their desperate patients. There is general agreement that clinical trials should be preceded by *some* basic research, while complete mechanistic explanations are not required. But this consensus on extremes leaves a large gray area open to argument. How much biological knowledge is necessary, before a clinical trial is warranted?

The vicissitudes of the first clinical trial using human embryonic stem cells (hESC) illustrate both the obstacles to and need for a principled answer.[203] In early 2008, Geron Corporation applied for approval of the US Food and Drug Administration (a legal requirement for clinical trials in the United States) to inject hESC into 8–10 otherwise healthy patients with severed spinal cords. The aim of this small Phase I trial was to assess safety and efficacy of hESC as treatment of severe spinal injury (see Figure 10.1). The FDA at first demurred, citing concerns that hESC might give rise to tumors after injection.[204] Then, in early 2009, approval was granted. The biomedical community's response was mixed. Many scientists thought the study premature, or questioned the choice of clinical target and patient population, or both. Bioethicist Arthur Caplan lambasted the Geron trial as "nuts, and hugely risky," arguing that more laboratory research on stem cells was needed (quoted in Regalado 2011). But other expert commentators were confident the trial would work, based on striking proofs-of-principle in model organisms. For many onlookers, the animal evidence, together with calls for cures by patient activist groups, seemed sufficient to justify the hESC study. No clear consensus emerged; instead, the biomedical community tacitly adopted a 'wait and see' approach.

As debate continued, with no clinical results available, the FDA again intervened, placing the trial on "clinical hold" in August 2009. The reason had nothing to do with human patients, who had not even been selected at the time. Rather, a high incidence of cysts was found in mice injected with cells similar to hESC. After further animal studies, which revealed no malignancy, the hold was released. But because very few patients met the study's criteria, the rate of data collection was extremely slow. The first human patient was injected with hESC in October 2010, and the study projected to run for three years. But in November 2011, Geron abruptly ended the trial, citing budgetary constraints in the lagging economy. No adverse effects on patients were reported – the trial was simply unfinished. Financial considerations appear to have trumped evidential ones. The Geron case raises a number of important questions about the role of industry in biomedicine. For my purposes, the key issue is lack of consensus about evidence relevant to the clinical trial. The trial's relation to laboratory experiments and their results was contested throughout, and, with no principled strategy for settling the issue, the biomedical community remained divided. The FDA's repeated vacillations reflect this wider ambivalence. Perhaps clear evidential arguments and strong community support could have held the ground against financial considerations. With more clinical trials for stem cells in the 'translational pipeline,' it is important to understand how laboratory research should bear on the decision to move forward.

Though the above is a general concern for translational fields, stem cell research presents particular difficulties. This is because the 'landscape of values' for stem cell biology is extremely complex. Stem cell scientists must adapt to political, ethical, and financial restrictions, making strategy and marketing central concerns alongside experimental work. Aspirations to scientific stardom, a revolution in medicine, profitable associations with industry, and so on, are easily conflated with promises of specific results. Wishful thinking on all sides skews assessment of risks and benefits. Terminology is also a problem: stem cells' open-ended potential is often communicated in hyperbolic terms, while ethical controversies are expressed in absolutes that connect poorly with experimental details. In this situation, it is difficult to determine what one's expectations should be, or what counts as distortion by non-scientific interests. Philosophy of science can help clarify these issues. The next section uses recent debates over "value-free science" to articulate a general framework for conceptualizing diverse values in translational research. Other clarifications are left for future work.

10.3 Value-free science

Debates about value-free science take place against a background of consensus. It is widely agreed that some values, such as empirical accuracy and consistency, are essential to good science. These values are classified as epistemic or cognitive.[205] It is also uncontroversial that other values, classified as ethical, political, or social (hereafter EPS values) often play a role in determining areas of inquiry and selecting questions to investigate, as well as guiding applications of results. EPS values include justice, communism, feminism, fairness, and nationalism. Economic and aesthetic values, though arguably not EPS values as such, are similarly implicated in scientific practice, and so are often treated as EPS values. Finally, it is well-established that EPS values have influenced theory choice in some paradigmatic cases of bad science, such as Lysenko's dismissal of Mendelian genetics, rejection of work by Jewish physicists under the Third Reich, and instances of blatant sexism and racism in the human sciences.[206] The main point of contention is the role of EPS values in determining the content of good scientific theories.

In a recent collection dedicated to this issue, editors Kincaid, Dupré and Wylie distinguish four dimensions of debate: (i) kinds of values involved in science; (ii) phases or contexts of science in which values are involved; (iii) whether value-involvement is essential or merely possible; and (iv) effect of value-involvement on scientific knowledge (Kincaid et al. 2007, 10–13). All four contested issues are closely related. Classification of values, contexts, essential features of scientific knowledge, and judgments of 'good-vs-bad' science all bear on one another, such that a full account of (i)–(iv) amounts to presumptively settling the debate at issue. The dichotomy of good-vs-bad science erects a boundary between science and other activities by providing a standard for normative assessment of scientific practices and their results. Such a standard also indicates what is essential to science, and what is not. Values deemed essential or conducive to good science are classified as epistemic, while values that are either not conducive to good science or associated with bad science are classified as non-epistemic.[207] Distinctions among kinds of values, their effects on scientific knowledge, whether those effects are essential, and judgments of good/bad science all mutually inform one another. If one of these distinctions is taken as primary, consequences for the others follow.

Classification of values as epistemic or non-epistemic tracks intuitions about good-vs-bad science. These intuitions vary considerably. They are, in turn, stabilized and clarified by philosophical accounts of the constitutive aim of science. The idea that science has a unique constitutive aim is reinforced by the simple, monolithic dichotomy of good-vs-bad science: good science is such as to satisfy this aim; bad science goes off-track. If science has a constitutive aim, then it should be guided by values necessary or conducive to achieving that aim. Epistemic values are those necessary for, or conducive to, the constitutive aim of science. The constitutive aim of science is thus entangled with the dichotomy of good-vs-bad science and with classification of values as epistemic or non-epistemic. To specify an epistemic criterion that unambiguously demarcates good from bad science is to specify *the* constitutive aim of science. For example, if good science yields significant truths about the natural world, then the constitutive aim of science is significant truths about the natural world. If good science yields truths likely to improve human flourishing, then the constitutive aim of science is truths that improve human flourishing. If good science yields accurate predictions of phenomena, then the constitutive aim of science is accurate prediction of phenomena. And so on. If science has multiple aims, then matters are more complicated. There is then no single demarcation of good-vs-bad science, but normative disunity and multiple dimensions of evaluation.

Given the (or a) constitutive aim of science, one can work out the epistemic values manifested in good science. Conversely, given a set of epistemic values for science, one can specify its constitutive aim, at least partially. Different conceptions of the constitutive aim of science yield different 'packages' of values. Values necessary or conducive to a unified system of significant truths about the natural world include consistency, empirical accuracy, and explanatory unification invoking natural kind terms. Values necessary or conducive to a collection of truths that can be used (separately or together) to improve human flourishing include empirical accuracy, diverse participation, and applicability to human needs. Debate over values in science is also debate over the constitutive aim of science.[208] But these debates cannot be settled a priori. Answers must be based on examination of scientific practices, past and present. This task is unmanageable, however, without some simplifying framework. Absent substantive background assumptions about the nature and aim of science, our practices are too diverse and their boundaries too contested for historical or contemporary observation to reveal constitutive aims or epistemic values.

The distinction between different phases or contexts of science (ii) is also implicated. The traditional distinction between contexts of discovery and justification serves to identify the epistemic core of science. This framework, in turn, supports a conception of the constitutive aim of science as manifest only in the context of justification. The context distinction also classifies as 'external' to science values that are properly involved in discovery and application, but not justification, of scientific claims. Though usually classified as external, EPS values are internal, or essential, to science if the privileged context of justification is understood broadly enough to include grounds for supporting particular standards for theory choice (Longino 1995). More generally, the distinction between internal and external scientific values presupposes a core of science, which may be analytically separated from other aspects of science, and from which all other values are properly excluded. In this sense, classification of values as epistemic or non-epistemic, essential or non-essential, depends on prior distinctions among contexts or phases of science.

If these three related distinctions are resolved, the fourth dimension of debate, effects of values on scientific knowledge, is thereby also settled. Various coherent accounts of values in science are possible, depending on how (i)–(iv) are specified. On the 'traditional package,' the epistemic core of science is assessment of theories by evidence, and the constitutive aim is formulation of theories that improve our understanding of the world. Values that guide theory assessment so as to further this constitutive aim qualify as epistemic. EPS values, in contrast, aim at a different ideal: the good society, or the world we should like to live in. So EPS values are prima facie irrelevant to the aim of science and, therefore, non-epistemic, external, and irrelevant for theory-choice. Insofar as such values do influence theory-choice, and thereby the content of science, then that content is not justified; proper assessment of theory by evidence, the core activity of science, has been subverted or distorted. EPS values, which concern how things ought to be, should not take the place of evidence for descriptions of how things are. This traditional account, a tidy, self-reinforcing package, epitomizes the value-free ideal for science.

Critics of the value-free ideal propose other, equally coherent, packages. For example, Lacey (2005) distinguishes between (i) epistemic and EPS values;[209] (ii) core and peripheral moments in science; and conceives science as aiming (iii–iv) to "generate and consolidate theories that express empirically-grounded and well-confirmed understanding of phenomena" (2005, 980). Epistemic values are defined as the characteristics of theories which, if accepted, further this constitutive aim; EPS values are excluded. Core moments in science are necessary means to that end, and include theory choice and selection of a strategy for inquiry. The latter involves methodological decisions as to "what constraints are to be put on theories that may be investigated (including what categories are to be deployed in them) and what kinds of empirical data are to be sought out and recorded" (ibid, 979). In good science, theory-choice is determined by epistemic values only, while non-epistemic EPS values have a legitimate role in strategy selection but not theory-choice. Decisions about the appropriate application of scientific knowledge are peripheral moments, and involve both epistemic and EPS values. Lacey's account, a more nuanced version of the value-free ideal, takes a 'theory-centric' constitutive aim of science as primary, and uses it to underwrite the distinction between epistemic and EPS values and an associated normative standard for 'good science.'

Many other accounts are conceivable. On one elegantly austere package, the constitutive aim of science is "significant truths," the epistemic core of science restricted to the evidential relation between theories and data, and epistemic (internal) values pertain exclusively to this relation. Less austere packages expand the epistemic core of science to include other aspects of theory choice, such that accurate prediction, explanatory unification, and fruitfulness qualify as epistemic. If the epistemic core of science is conceived at the community level (as in Longino 2002), then at least some values bearing on social relations are also epistemic. Explicitly political packages are also conceivable. If the constitutive aim of science is to improve human flourishing, then ethical values are properly involved in the epistemic core. And similarly for more specific projects aimed at human equality, such as democracy and feminism (Kitcher 2010; Kourany 2010). EPS values may be classified as internal or external to science, depending on which internally consistent package one prefers. There is no absolute or a priori standard that privileges one account over others. So, in general, pluralism about values in science prevails. However, in particular cases, the interrelated questions of context, aims, and values are more tractable. Case studies of human sexuality and aggression (Longino 1995), the impact of divorce (Anderson 1995), risk assessment (Douglas 2009), and transgenic crops (Lacey 2005) have illuminated the roles and kinds of values at work in these fields. Stem cell biology can also be examined in terms of distinctions (i)–(iv).

10.4 Values in stem cell biology

Of these distinctions, the constitutive goal is most easily discerned. Of course, the specific goals of stem cell research are very diverse, targeting a wide range of pathologies: heart attacks, blindness, cancer, neurodegenerative disease, paralysis, diabetes, complications arising from organ transplantation, and many others. But all share the same basic idea: harness stem cell capacities for therapy. Stem cell biology does not aim at 'pure knowledge,' but at knowledge useful for therapeutic ends. Its goal is thus both epistemic and ethical. Previous chapters focus on the epistemic aspect: knowledge of cell developmental mechanisms, supported by experimental evidence and supporting model-based predictions in turn. This knowledge, in the form of mechanistic explanations, goes hand-in-hand with control of stem cells' distinctive capacities. This 'controlling' project has deep ties to medical practice and principles. Stem cell research aims at explanations that indicate how to harness regenerative cells for therapy, as renewable sources of healthy tissue. By directing cells to develop along pathways of our choosing, it is hoped that clinicians will be able to repair of a wide variety of damaged organs and tissues: severed spinal cords, damaged heart muscle, leukemic immune systems, diabetic pancreas, and many more. To bring about these medical advances (if they are possible at all), we need to know how stem cells work within organisms – in particular, in humans. Knowledge of stem cells that does not contribute to this clinical aim is not essential.

The clinical aim of stem cell research is grounded on principles of medicine, which include reduced harm and suffering and increased length and quality of life. The nature and formulation of these principles are subjects of much debate, but their ethical aspect is not: principles of medicine are ethical principles. In this way, a subset of ethical values (hereafter 'clinical values') is implicated in the constitutive goal of stem cell research. This has consequences for the other distinctions noted above. 'Good stem cell research' satisfies both epistemic and ethical constraints. Knowledge produced by stem cell research should also be essential or conducive to therapies that conform to principles of medicine and which are, in this sense, ethically, as well as epistemically, good. It follows that there is no core of stem cell science from which clinical values are properly excluded. Norms for experiment, explanation, and modeling in this field are informed by these values. Therefore, stem cell biology does not conform to the ideal of value-free science. However, not all EPS values are implicated in the principles of medicine. Those that are not, including political and economic interests, can be excluded from the core of stem cell research. The latter are, of course, relevant for many aspects of translational stem cell research. But the epistemic core of stem cell biology, which includes experiment, explanation and modeling, excludes EPS values other than clinical values implicated in ethical principles of medicine.

This result only follows if the therapeutic goal of stem cell biology is *constitutive*. There are at least three reasons to accept this premise. First, the clinical goal is fundamental in a historical sense, to both branches of stem cell biology (Chapters 7 and 8). The field assumed its current configuration only after hESC cell lines were constructed and explicitly linked to therapeutic aims (Thomson et al. 1998). Before 1998, pluripotent stem cells were a technology for genetic intervention, not the focus of a new research field in their own right. Their association with an explicit therapeutic goal unified and galvanized a new research community. Studies of tissue-specific stem cells, the other major branch, emerged in close association with cancer therapy. The first stem cell assay was developed by cancer researchers (Till and McCulloch 1961), while bone marrow transplantation therapies both spurred and benefited from decades of blood stem cell research in mice and humans.

Second, the clinical aim is crucial in maintaining the field's current organization. Without this shared focus, the field could easily fragment into a loose patchwork of techniques and models distributed across biomedical disciplines – as was the case for pluripotency research before 1998. The widely-shared clinical goal unifies stem cell biology as a community, facilitating collaboration between quite disparate experimental fields and disciplines. It is therefore implicated in mechanistic explanations of development, which are constructed by experimenting communities. The more inclusive the experimenting community, and the more diverse its models and experimental systems, the more robust and comprehensive explanations it can produce (Chapter 7). For example, mechanistic explanations for reprogramming require extending the experimenting community of cell reprogrammers to include systems biologists (Chapter 9). Similarly, translational research aimed at testing the cancer stem cell hypothesis requires close coordination between surgeons, laboratory researchers, and clinicians (Chapter 8). Increasingly, stem cell researchers are attempting to coordinate their research on a global scale, with large-scale standardization efforts designed to increase efficiency in collaborations (e.g. Adewumi et al. 2007). The shared clinical goal provides a stable focus for these and other efforts, spurring integration of results from different experiments to produce community-level explanations.

Third, the clinical goal influences "gold standards" for individual stem cell experiments. The gold standard for pluripotency research is to create a cell line with the same molecular features and cellular capacities as those of 'canonical' embryonic stem cell lines: unlimited cell division (self-renewal) and pluripotency, the potential to differentiate into all cell types of the adult organism (Thomson et al. 1998, 1145). Certain traits of hESC make them exemplary models for pluripotency research: wide differentiation potential, low tumor formation, minimal manipulation, and high reproductive rate (Chapter 7). But these traits are exemplary only relative to the field's clinical goal. The gold standard for this branch of stem cell research is, in effect, an optimal source of cells for therapy. Of all stem cells characterized to date, hESC are the closest approximation of this standard, capable of rapidly producing all relevant tissues, closely resembling normal human cells, and minimizing risks of cancer or infection.

Tissue-specific stem cell research shows the same pattern. The gold standard in this branch is to extract cells from an organ or tissue, select the stem cells in the tissue using surface markers, immediately transplant one cell into a host animal, and observe self-renewal and reconstitution of the relevant tissue, organ, or cell type for the lifespan of the host (Melton and Cowan 2009, xxvii). The exemplar that most closely approximates this standard is the blood stem cell (HSC) method, originally developed in mice, but soon after applied to humans (Chapter 8). Human HSC transitioned to the clinic decades ago, and are today routinely extracted from bone marrow, circulating blood, or umbilical cords. They remain the only stem cells in wide clinical use.[210] As with hESC, the exemplary status of HSC is relative to clinical aims.

To sum up: the clinical goal of stem cell research is constitutive because it is historically fundamental, necessary to the field's continued existence and epistemic improvement, and helps determine prevailing standards for experiments and explanation. The latter have normative consequences, many of which were discussed in earlier chapters. Stem cell biology's constitutive goal grounds norms of diversity among model systems, integration of experimental results, and expansion of experimenting communities. Clinical values thus take on a unifying role not unlike that traditionally assigned to fundamental theory. Stem cell explanations should help us safely manipulate stem cell capacities within *human bodies*. These explanations are therefore constrained to be both robust and detailed, so as to support reasonable predictions about outcomes of experimental treatments. Reductive accounts that focus on one or a few "master molecules" are not equal to this challenge. Successful explanations must include the system-level of whole, functioning organisms. In this way, the clinical goal supports increasing convergence with systems biology (Chapter 9). However, this aim also skews modeling practices toward concrete material models, and away from more abstract constructions favored by systems theorists (Chapters 7 and 8). Clinically-useful explanations should eventually erase the distinction between concrete stem cell models and their primary target: cells developing in human bodies.

At still a finer grain, the clinical goal of stem cell research furnishes a standard for evaluating models that inform substantive stem cell concepts. These abstract models implicate users' purposes in representation relations and choice of parameters (Chapter 2). Those purposes include the clinical goal, which thereby constrains acceptable stem cell concepts. For example, many cell lineage models aim to represent single cells and links between stages of cell development; features highly relevant to the clinical goal of stem cell research. A challenge that remains to be met, however, is imposing a clear distinction between features of stem cell models that represent how *cells* work irrespective of human intervention, and features that reflect our manipulations – how *we* work to identify and characterize cells. Articulating this version of the subject/object distinction will be crucial for success in translational stem cell research going forward.

The clinical goal of stem cell biology also suggests specific directions for future progress. The trend in both branches is toward expansion of experimenting communities, which produces increasingly unified and robust explanations. In the 1980s, this norm of 'explanatory uni-fication' proved crucial in the search for HSC, which converged on the common stem of diverse lineage pathways (Chapter 8). Today, it spurs synergistic working out of similarities and differences in related experimental systems, in tissue-specific and pluripotent branches. In each, diversity of model systems increases as the field moves forward. Though much has been made of iPSC as a substitute for hESC, bypass-ing the need to use human embryos in research, in practice the two are used together to study cell differentiation. Continuing this trend, the two branches of stem cell research should eventually merge into a single experimenting community. The search for commonalities among different tissue-specific stem cells, together with studies of "plasticity," should reveal convergent developmental origins. On the other side, directed differentiation of pluripotent stem cells should reveal 'down-stream' developmental possibilities. The two approaches are poised to merge in a more comprehensive and clinically-useful account of cell development. If this line of reasoning is correct, then the 'adult-vs-embryonic' dichotomy, which represents the two branches as alterna-tives, is misconceived.

10.5 Conclusion

The above result illustrates one way that philosophy of science can bear on stem cell research policy. If funding priorities are set in terms of a presumed opposition between the two branches of stem cell research, progress in both will be impeded. More subtly, the idea that the two branches are competing alternatives works against the development of comprehensive explanations of development, and thus against full realization of the field's clinical aims. In examining this goal, with its dual epistemic and ethical aspects, the present study joins other projects in socially-relevant philosophy of science (Fehr and Plaisance 2010). This chapter, clarifying the role of ethical values at the core of stem cell science, is only a first step. Much remains for future work. I will close with a few suggestions. One significant project to which philosophers can contribute is translational stem cell research policy. The clinical goal, and ethical values underlying it, provides a clear starting point for discussions between experts and non-experts about the direction (and funding) of stem cell science, and a basis for translational researchers' accountability. Justification of particular stem cell research projects should include an explanation of how the experiment fits with the field's constitutive goal. Such explanations should be accessible to non-experts, and suggest standards for judging success or failure that can be made clear to all interested parties.

Another promising topic for investigation is the structure of collaboration in translational stem cell research. The complexity of hoped-for clinical interventions requires interdisciplinary teams, which need shared standards and reliable indicators of trustworthiness to work effectively. The 'hope and hype' surrounding stem cell research are notorious. Stem cell biologists stress the need for inclusiveness, "moving the field forward together," but their perspectives on its boundaries and internal organization are necessarily partial (Simmons 2007). Intense collaboration can foster insularity and reinforce shared biases within a community, subverting the goals and norms discussed earlier. Philosophers of science could partner with historians and social scientists to make valuable policy recommendations. More traditional theories of evidence and error could help manage the uncertainties of clinical trials and inform guidelines for reasonable expectations about research outcomes. Social epistemic accounts could help to formulate norms of transparency and information-sharing, for predictions about short-term clinical results, and counteract short-sighted ideas that the best science is that which quickly produces useful (or lucrative) results. All these projects fit with a turn toward socially-relevant philosophy of science.

To conclude the present study: a view toward the clinic pervasively influences stem cell biology. Rather than distorting its results, these clinical values guide stem cell research and provide a basis for normative assessment of its models, experiments, and explanations. So stem cell biology is not a "self-vindicating" laboratory science, with theories, technical apparatus, and methods of data-analysis mutually adjusted to produce a self-reinforcing system that is "essentially irrefutable" (Hacking 1992, 30). Its goals extend beyond a stable, coherent system for generating well-confirmed experimental results. It aims, instead, at explanations that suggest ways of manipulating cells to treat a variety of pathological conditions. A unified mechanistic account of cell development will help us build effectively on present clinical achievements, notably those of blood stem cells, as well as recognizing their limitations.

The quotations at the start of this book speak, from opposite ends of history, to the hopes for regenerative medicine. The more recent is an optimistic slogan for twenty-first-century biomedicine: "biology is not destiny but opportunity" (Rose 2007, 51). But opportunity for whom? If stem cell science achieves its goals, we will literally be able to re-make our bodies. The prospect is inspiring, but also unsettling. Stem cell biology's dual aims of understanding and control of cell development are easily confused with total, hubristic domination over life itself. To forestall these concerns, many commentators stress the need for explicit, universal guidelines, formulated by international groups of scientists in ongoing dialogue with regulators, clinicians, and the wider public. I have argued that philosophers of science should also participate in these discussions. But the deeper point is that collaborative efforts should not be seen as external constraints, which must be imposed to trammel the unbridled egoism of stem cell research. The clinical goal of the field pervades stem cell experiments and models, even in the laboratory. Making explicit the ways it does so will help the stem cell field develop responsibly and with accountability to those who support it. The control to which stem cell researchers aspire is not absolute domination over life itself. Stem cell experiments could not provide this in any case, as stem cells are always poised to elude our grasp. But the purpose of stem cell research is more modest, qualified, and closer to home. What is sought, in laboratories and increasingly in clinics, is control for therapy on the scale of human lives.

Notes

1. Statistics on gut from Potten (1983).
2. Most stem cell research focuses on animals rather than plants, though both undergo development (and stem cells in *Arabidopsis* are an important research area). Accordingly, this book deals, for the most part, with stem cells in animals. Any full explanation of development, however, must also extend to plants.
3. Some stem cells (of blood and skin respectively) are currently used to treat cancers and burns. But the therapeutic potential of stem cells goes far beyond this.
4. *Cf.* Clarke and Fujimura (1993).
5. Examples include: Maienschein (2003), Fox (2006), Landecker (2007), Monroe et al. (2008), Gottweis et al. (2009), and Maehle (2011).
6. The meaning of 'stem cell' is, of course, not fixed, but varies over time and across disciplines. For histories see Maienschein (2003), Shostak (2006), Ramalho-Santos and Willenbring (2007), and Maehle (2011).
7. Sources: (a) "Stemness: definitions, criteria, and standards," Melton and Cowan (2009, xxiv) in: *Essentials of stem cell biology*, 2nd edn. Preface (Lanza et al., eds); (b) "On the origin of the term 'stem cell'" Ramelho-Santos and Willenbring (2007: 35); first issue of *Cell Stem Cell*, the journal associated with the International Society for Stem Cell Research (est. 2002); (c) National Institutes of Health, US, "Stem cell basics: what are stem cells?" (2009); (d) European Stem Cell Network, "Stem cell glossary" (2008).
8. See Landecker (2007) for the history of cell culture and its broader significance.
9. These experiments are discussed in more detail in the following sections and in later chapters.
10. See later sections in this chapter and Chapter 8.
11. For more information on these topics, see the stem cell information pages at http://stemcells.nih.gov/info and http://isscr.org/faq
12. See §2.1.2.
13. Biologists often refer to 'mother' and 'daughter' cells. I avoid this terminology in this book. Gendered language is laden with myriad assumptions and biases associated with gender divisions in society; cell biology is complex enough without such baggage. Women in science (and philosophy) who struggle with obstacles created by entrenched assumptions and biases have good reason to be suspicious of gendered language, even if it seems harmless. Gender-neutral terms for cellular processes are used throughout.
14. One rationale for this convention is that offspring pairs are more easily compared with one another than cross-generation pairs. A parent cell divides to produce two offspring cells; when they exist, the parent does not and vice versa. So side-by-side comparison of parent and offspring cells is impossible, while side-by-side comparison of offspring cells with one another is often feasible.

15. Self-renewal at the population level requires only that some cells in the offspring population have the same character values as some cells in the parent population.
16. Another much-debated feature of models, their relation to *fiction*, is put aside here (though see Chapter 7, for related discussion).
17. In practice, this third set is often empty.
18. See, respectively, Van Fraassen (1989), French and Ladyman (1999), and Suárez and Cartwright (2008).
19. Giere (1988, 2004). This formulation allows for different characterizations of the domain of Y and the nature of representation. For example, the domain of Y may be conceived as aspects of the real world (Giere 2004), empirical phenomena, or empirical structures (Hughes 1997). The representation relation may be conceived as isomorphism (Van Fraassen 1989), similarity (Giere 1988), or inferential connection (Suárez 2004).
20. Cartwright et al. (1995), Morgan and Morrison (1999), Cat (2005), Suárez and Cartwright (2008).
21. Alternatively, τ decreases at each level as defined by DF.
22. There are then three cases to consider: capacity for SR differs and capacity for DF is the same, capacity for DF differs and capacity for SR is the same, or both capacities differ.
23. Duration of self-renewal also decreases progressively with levels. This is stipulated by (M8), but also required by any realistic causal interpretation of the model. Maintenance of a cell reproductive hierarchy over time requires longer duration of self-renewal for higher than lower levels. Otherwise, lower levels would simply replace higher ones; the hierarchy would not persist over the time interval in question.
24. *Cf.* Keating and Cambrosio (2003).
25. Choice of measurement also influences the concept of self-renewal. The capacity to self-renew for ten cell cycles is the capacity to undergo ten self-renewing divisions, whether this takes a week or several decades. But the capacity to self-renew for ten years means that cell divisions occurring in that interval yield offspring with traits like the parent, maintaining cells with those traits in the organism's bodily economy.
26. See Maehle (2011) for early uses of the term 'stem cell' and its cognates.
27. Though no comprehensive history of stem cell biology has yet been written, several recent studies trace particular aspects, including embryology (Maienschein 2003), reproductive technology and cloning (Fox 2006), cell culture (Landecker 2007), radiation biology, and hematology (Kraft 2009). Maehle (2011), though brief, makes an admirable start on such a history.
28. Some features of the original iPSC experiments, later found unnecessary, are here omitted for simplicity. Reprogramming experiments are further discussed in Part II (esp. Chapter 6).
29. These were: cell surface molecules, activity and expression of specific proteins, expression of specific genes, global gene expression, and histone modification of specific genes (Takahashi and Yamanaka 2006, Takahashi et al. 2007).
30. The original hESC protocol included a layer of cultured 'feeder cells' that supported the ICM-derived cells, both physically and by producing growth factors.

31. The key character values were: rapid division in culture, homogeneous appearance, lack of specialized traits, flat round shape, large nuclei surrounded by correlatively thin cytoplasm, and prominent nucleoli. See Chapter 7 for more on EC.

32. These characters were: chromosome number and appearance, telomerase activity, and cell surface molecules.

33. See Chapter 8.

34. iPSC and ESC also use comparisons to other stem cells, but this is not a feature of the HSC method.

35. In practice, 160 cell doublings is a popular standard (Melton and Cowan 2009, xxiv).

36. Before 2006, there was a strong inverse correlation between age of the source organism (from time of conception) and developmental potential. For normal development, the correlation still holds up well. But stem cell methods involve departures from normal developmental pathways. iPSC from adult organisms have developmental potential similar to ESC.

37. See Chapter 4.

38. This section closely follows Price (1995, esp. 390–392). All quotations in this section refer to this work unless otherwise specified. For more on Price's evolutionary theorizing, see Price (1970, 1972), Frank (1995), and Okasha (2006).

39. Price intended the model as a step toward a "general selection theory and general selection mathematics," which would subsume Darwinian natural selection as a special case (p. 389). Price took his own life before completing the project, which remains unfulfilled.

40. This formulation follows Okasha (2006). The covariance is of fitness and character value. The expectation is of the product of fitness and the difference in character value between parent and offspring. ΔX is the change in average character value from P to P', $\bar{\omega}$ is average fitness in P.

41. *Cf.* Price (1995, 391).

42. Interestingly, however, Price's general selection model and the abstract stem cell model converge as a special case (for both). The two coincide if the former is applied to cells undergoing division and development, and latter is elaborated with R2 and includes both statistical properties of cell populations over time and individual cells' parent–offspring relations.

43. Stem cell biology encompasses many kinds of experiment other than these exemplary methods. For brevity, I refer to experiments that aim to isolate and characterize stem cells as 'stem cell experiments.' But the moniker is for convenience; it should not be taken as describing *all* experiments involving stem cells.

44. Genotype or strain is also frequently used.

45. For references, see §2.5.

46. The three most common treatments enabling measurement are: (i) chemical fixatives and stains to visualize cell morphology and intracellular structures; (ii) fluorescently-tagged antibodies that bind specific surface molecules, labeling cells with 'immunofluorescence;' and (iii) breaking cells apart to isolate specific molecular components: DNA, RNA, or protein.

47. Other in vivo experiments test for production of a whole organism or all its tissues, including the germline, from transferred cells. But these experiments

are not performed in humans; *m*ESC, but not hESC, pluripotency is tested in vivo.

48. Mice used in laboratory experiments are highly inbred, such that genetic variation within a 'strain' is extremely low. mHSC are characterized by strain, as well as species.

49. Mouse, but not human, ESC can be transferred to new cultures as single cells. But such transfers are not always done. In mHSC experiments, as few as 40 cells of a sorted subpopulation may be 'diluted' with unsorted bone marrow cells (Spangrude et al. 1988).

50. Thresholds proposed by the International Stem Cell Initiative (Adewumi et al. 2007: 812–813).

51. g_1 may be qualitative or quantitative, all-or-nothing, or a matter of degree.

52. A successful experiment is one in which well-calibrated methods work as expected, with all necessary controls, and yield a clear and credible result. We can further suppose this result (e_g) is *reproducible*, i.e. any cells manipulated and measured in this particular sequence (including organismal source) yield this result. These assumptions bracket other evidential concerns about experimental data, which are not problems for stem cell biology in particular.

53. Alternative quantitative definitions use probability or likelihood ratios (Sober 2008). All three give the same verdict on the stem cell case.

54. It follows that $Pr(H_p|e_g) \approx 1$:

$$Pr(H_p \mid e_g) = \frac{Pr(e_g \mid H_p) \times Pr(H_p)}{Pr(e_g \mid H_p) \times Pr(H_p) + Pr(e_g \mid \neg H_p) \times Pr(\neg H_p)}$$
$$\approx \frac{Pr(e_g \mid H_p) \times Pr(H_p)}{Pr(e_g \mid H_p) \times Pr(H_p)} = 1$$

If $Pr(H_p)$ is rather low, then the data confirms H_p to a high degree.

55.

$$Pr(H_s \mid e_g) = \frac{Pr(e_g \mid H_s) \times Pr(H_s)}{Pr(e_g)} \approx 1 \times Pr(H_s),$$

and degree of confirmation is defined as $Pr(H_s|e_g) - Pr(H_s)$.

56. This is distinct from the question of representative sampling of a cell population P. The distribution of G-values in P is also relevant for determining how large a sample is needed. Because most samples include many cells ($\geq 10^3$) adequate sample size is usually assumed. But this cannot be demonstrated without information about the variance in G in P. The problem for H_s is more serious, however. Even if sample size is adequate, e_g is not good evidence for H_s if P is heterogeneous with respect to G.

57. 'Gold standards' are from the 2009 edition of *Fundamentals of Stem Cell Biology* (Lanza et al. 2009) and the International Stem Cell Initiative's 2007 characterization of hESC lines (Adewumi et al. 2007).

58. Recent examples (circa spring 2011) include: quantitative measurement of RNA transcripts in a single cell (single-cell reverse transcription-polymerase chain reaction), live imaging to track single cells in regenerating tissues

(confocal and two-photon microscopy), and refinements to flow cytometry increasing the number of simultaneously measurable characters of a single cell (from special issues of *Nature Reviews Genetics*, April 2011, and *Nature Cell Biology*, May 2011).

59. Terms from Loeffler and Potten (1997).
60. Details in Vogel et al. (1969).
61. Twenty-five per experiment, with multiple replicate experiments.
62. Shared chromosomal aberrations within a spleen colony provided the most persuasive such evidence; serial dilution of cell preparations and tracking the form of survival curves were also used.
63. Values ranged over 1–2 orders of magnitude for a coefficient of variation of 2.0 (Till et al. 1964, 32).
64. See Loeffler and Potten (1997) for additional examples.
65. This is always the case for differentiation; pluripotency methods, however, impose self-renewal at the outset of the experiment (see §2.6).
66. This brief treatment is intended to clarify the views put forward by scientists, not to subject these views to deeper metaphysical analysis.
67. Zipori does not offer this summary himself; I take (i)–(iv) to be his core ideas, distilled from various sources.
68. This chapter follows Zipori in using 'stemness' and 'stem state' interchangeably. For example, a stem state is defined as "having many potential outcomes but no specialization" (2004, 876).
69. Zipori and other proponents of stemness recognize that many stem cells do self-renew, but consider this attribute non-essential to the stem state.
70. Or, at least, wide differentiation potential relative to cell types in the adult organism.
71. See §2.4.3A.
72. In this way, the collinear model helps underwrite the "adult-vs-embryonic" dichotomy.
73. For example: "Of late, the traditional notion of stem cells as a clearly defined class of intrinsically stable biological objects that can be isolated and purified, has begun to give way to a view of 'stem-ness' as temporary, shifting and evanescent ... In the course of these shifts and turns, the 'stem cell' becomes a fleeting, ephemeral and mythical entity" (Brown et al. 2006, 339–343). Similar claims can be found in Martin et al. (2008), and Moreira and Palladino (2005).
74. The stemness account here goes beyond the uncertainty principle, which concerns prospective isolation of single stem cells. Cell compartments and populations are also "discrete cellular entities."
75. Realists also require that successful theories be true, or approximately true, or that we have good evidence that they are approximately true. Many stem cell models and hypotheses conform to realist standards.
76. This is one motivation for shifting philosophical focus from theories to models.
77. The apex of the hierarchy is the fertilized egg of a multicellular organism. However, the fertilized egg is not a stem cell, as it does not both self-renew and give rise to more differentiated cells within a given time interval. More precisely, it does both, but at different time-scales, so the abstract model's definition is not satisfied.

78. Another way to put the point is that experiments indicating adult stem cell plasticity in certain (highly artificial) contexts are taken to support a general account of "stemness" that is independent of all such contexts, with stem cells from various sources "interchangeable" in their wide developmental potential. This places plasticity claims (and the stemness view) beyond the reach of experimental test. Note that plasticity claims are not thereby disconfirmed.

79. *Contra* Brown et al. (2006, 339).

80. Zipori (2009) attributes self-renewal entirely to the niche, and differentiation potential to the internal cell state.

81. Craver (2007) and Illari and Williamson (2010) make this point very clearly.

82. This term is from Weber (2005).

83. The reference to "causal laws" is later replaced by "invariant, change-relating generalizations" (Glennan 2002). See section 3 for discussion of the latter term.

84. Details in Yamashita et al. (2003), Sheng et al. (2009), and Cherry and Matunis (2010).

85. For 'Janus kinase signal transducer and activator of transcription.'

86. The term 'mechanism' is also sometimes used to describe non-hierarchical causal chains; "etiological mechanisms," in Craver's (2007) terminology. Here, "mechanism" refers only to hierarchical (or "constitutive") mechanisms.

87. See Craver (2007, Chapter 5) for a persuasive, systematic defense of this claim.

88. See Bechtel (2006), and Bechtel and Richardson (1993/2010).

89. Compare the general sketch of asymmetric cell division in Figure 2.1 with the GSC mechanism depicted in Figure 5.1. Both represent asymmetric cell division, but the details that are crucial for the latter are absent from the former.

90. This formulation follows the definition of Machamer et al. The term 'φ-ing x's' is used by both Craver and colleagues (see Figure 5.2) and philosophers of social action (see note 106).

91. See Salmon (1989) and Woodward (2003) for critical discussion of the covering-law theory.

92. This way of framing the problem follows Schaffner (1993).

93. See Craver (2007) for other examples illustrating this point.

94. Woodward's full analysis is more elaborate. But this simplified treatment suffices for my purposes here.

95. Woodward, however, associates explanatory power with invariant, or stable, generalizations, retaining the core idea of the covering-law theory (see §5.7).

96. Though Craver restricts his claims to neuroscience, his causal-mechanical account extends to experimental biology more generally.

97. The concept of manipulation here is Woodward's: a controlled and predictable change, which can be represented by a dependency relation between values of variables.

98. "Non-aggregative" means that altering the number, properties, or spatio-temporal arrangement of parts makes a difference to biological mechanisms' overall working. "Decomposability" is a way of classifying complex systems by the degree to which their behavior is predictable from aggregate behavior

of isolated components: decomposable systems are predictable, nearly-decomposable systems approximately so, and non-decomposable systems not at all.

99. For cyclic or continuous processes, there is no outcome distinct from the process itself: blood circulates, biochemical networks metabolize, organisms age.

100. Important examples are Bechtel and Richardson (1993), Darden (2006), and Bechtel (2006).

101. Though Craver does not conflate the constitution relation with causation, he articulates and defends the multilevel structure of MEx by "close analogy" with and "extension of causal relevance" (ibid, 152, 162).

102. 'In principle' here means that such an experiment is conceivable not that it can actually be performed.

103. Woodward (2002) does not distinguish between the modularity theses proposed here, though (as I interpret him) all but *Mod-4* follow from the manipulability theory.

104. This formulation, and those below, are not Woodward's own, though they capture what I take to be the main ideas of his account of modularity in MEx (Woodward 2002). Modularity in his theory as a whole has a rather different significance.

105. Decoupled from realist assumptions, *Mod-4* still constrains MEx. Craver's account presupposes realism about causes.

106. This term is from social action theory where actions of two or more agents, such as walking together, dancing a tango, or building a cathedral, and associated intentions, are variously referred to as 'joint,' 'collective,' 'social,' or 'cooperative.'

107. The lock and key model is owing to Emil Fischer, who further hypothesized that each enzyme has a unique substrate to which it binds to catalyze a unique chemical reaction. Though his hypothesis is long-since disconfirmed, the lock-and-key motif remains a prominent feature of experimental biology (see Chapter 6 for further discussion).

108. For example: components of mechanism S x_1 and x_2 jointly ϕ (x_1x_2 ϕ-ing), and if x_1 and x_2 did not form complex x_1x_2 in S, then x_1 and x_2 would not ϕ in S.

109. For simplicity, only the two-component case is presented. Condition (ii) of J3 is redundant in the two-component case, but not for complexes with three or more components.

110. See, for example, Glennan (1996, 55–56). The MDC definition (§5.2) generalizes this linear view.

111. See Carlson (1966, 259), Beurton et al. (2000), Moss (2003), Burian (2005), and Griffiths and Stotz (2007).

112. See Moss (2003) for incisive critique of (G').

113. After Burian (2005).

114. See §3.3 and §4.4 for more on gene expression.

115. Paraphrased from Waters (2007, 566–567). The final requirement is further explicated by several conditions pertaining to specificity, which is discussed later in this section. A cause X that satisfies all but this last condition is *an*, but not *the*, actual difference-maker for Y in P.

116. *Cf.* Waters 2007 (19).

117. Though Woodward (2010) does not explicitly argue for DNA privilege, his specificity concepts support such arguments.
118. Causal relations that are PI-specific are (approximately) one–one specific relative to some range of effects. However, a causal relation may be one–one specific, but non-PI-specific (Woodward 2010). So the two concepts are distinct.
119. Reviewed in Maherali and Hochedlinger (2008), Hochedlinger and Plath (2009), and Stadtfeld and Hochedlinger (2010).
120. See Chapter 7 for further discussion.
121. See Chapter 7 for relations among tumors, embryos, and ESC.
122. For a good non-technical description, see Carroll (2005).
123. Human genes are capitalized as well (*NANOG*).
124. Examples in Takahashi and Yamanaka (2006, 674), Yamanaka (2009, 51), Stadtfeld and Hochedlinger (2010, 2240), and many others.
125. These "cloning" techniques were later applied to mammals, culminating in Dolly the sheep.
126. The difference is conceptual rather than methodological, as cloning experiments can be conceived as revealing extracellular factors' effects on cells and TF as traversing from cytoplasm to nucleus to induce pluripotency.
127. Waddington coined the term "epigenetic" as a syncretization of "epigenesis" (the idea that embryonic development involves gradual formation of new parts, rather than growth of pre-existing forms) and "genetics." Today, "epigenetic" typically refers to molecular mechanisms producing changes in gene expression that do *not* involve DNA sequence changes. Several commentators (see Van Speybock et al. 2002, and Jablonka and Lamb 2005) have pointed out problems with this catchall category and confusions that arise from conflating different senses of "epigenetic." To minimize these problems, I use the term sparingly in what follows.
128. *Organisers and Genes* (1940), *An Introduction to Modern Genetics* (1939, 182–184), *Principles of Embryology* (1956), and *The Strategy of the Genes* (1957).
129. Waddington identifies the developing entity variously as a fertilized egg, a region of egg cytoplasm, and part of developing organism (a "piece of developing tissue;" 1940, 45).
130. Haraway (1976, 115–121).
131. Waddington (1940, 45, 92).
132. Chapter 2, Gilbert (1991).
133. These co-options of Waddington's model occur in journals (*Nature, Development,* and *Cell Stem Cell;* see references to figures in this section), workshops (Ichida et al. 2010), and international conferences (J. Thomson, 2010 ISSCR plenary address).
134. Chapter 5 critiques this modularity assumption, among others. Though the arguments are not repeated here, deviations from modularity present a further problem for the regulatory genome account.
135. Davidson 2006 (122).
136. Human embryonic stem cells are also the source of moral controversy about stem cell research. As much has already been written about these debates, I do not rehearse the arguments here. However, results of this chapter are relevant to ethical debate over hESC (see §7.6).

137. Influential case studies of model organisms include: Kimmelman (1993) on maize, Mitman and Fausto-Sterling (1993) on the flatworm *Planaria*, Kohler (1994) on *D. melanogaster*, Ankeny (1997) on *C. elegans*, Creager (2002) on tobacco mosaic virus, and Löwy and Gaudillère (1998), and Rader (2004) on *M. musculis*. Important collections dealing with model organisms include Clarke and Fujimura (1993), and Creager et al. (2007). For studies of three-dimensional material models other than organisms, see de Chadarevian and Hopwood (2004).

138. In some cases, evolutionary arguments support the inferences in question. However, well-supported evolutionary arguments are not currently available for most developmental phenomena. So, for now, this inferential path is blocked in stem cell research.

139. See Chapter 2.

140. Rheinberger refers to such mutual re-shapings as "conjunctures."

141. Unfortunately, this study conflates human embryos with hESC, making it less useful for understanding stem cell research.

142. Though tissue-specific stem cell research also uses cell culture methods, stem cells in this branch do not fit the model organism profile (see §7.4 and Chapter 8).

143. See Landecker (2007) for more discussion.

144. Argued in §7.5.

145. Readers who endorse a concept of organism with requirements that stem cell lines do not meet may substitute the clunkier term "organic parts of model systems."

146. Stem cell lines also resemble cancers; indeed, the linkage is even deeper (see §7.5.1).

147. See references in §7.2.

148. Similar tumors containing only differentiated cells are benign "teratomas," though this terminology (like designations of cell potency) was standardized later.

149. One embryoid body, which appeared within a human teratocarcinoma, was used as an atlas of human embryonic tissue!

150. These include humans, but, because of ethical constraints, knockouts are limited to mice.

151. *Cf.* Andrews (2002): "most of the use of mouse ES cell technology over the past 20 years has been directed toward production of transgenic mice, and not for answering questions of fundamental cell biology pertinent to ES cells *per se*" (412).

152. *Contra* Robert (2005).

153. These mice are another important model organism, engineered to be immuno-deficient to prevent graft rejection.

154. The "gold standard" for pluripotency in mESC, blastocyst injection producing chimeric mice with ESC-derived germlines, is ethically impermissible in humans. So teratoma formation spanning all three germ layers is the most stringent criterion for hESC pluripotency.

155. This approach has succeeded with other mammals, however. Fox (2006) includes an excellent discussion of the scandal.

156. For historical background, see Fagan (2007, 2010) and Kraft (2009).

157. See Chapters 2, 3, and 7.

158. This phrase is from Steinle. This paragraph and the next closely follow his (2002). Instrumentation associated with systems biology is also associated with exploratory experimentation in another sense (see Chapter 9).
159. See Rheinberger (1997) and Galison (1987), respectively, for examples.
160. See §3.3.
161. As in pluripotency research (Chapter 7), HSC methods were pioneered in mice and then extended to humans.
162. See Keating and Cambrosio (2003) for details.
163. The phenomenon does not occur in humans, indicating species-specific differences in blood and immune cell development.
164. The key assumptions are that ~10% of donor HSC form colonies and that their differentiation potential varies stochastically (see Chapter 3 for details).
165. Each also included a distinctive purification step: density centrifugation for the Rijswijk report and an immunomagnetic column for the Stanford report.
166. More precisely: the Rijswijk conclusion was that HSC have high density (as measured by ultracentrifugation), bind WGA, express the major histocompatibility antigen H-2k, and appear undifferentiated by light microscopy. The Stanford conclusion was that mouse HSC do not express surface markers of the main differentiated blood cell lineages (B⁻M⁻G⁻), and do express low levels of Thy-1 (Thy-1^{lo}) and high levels of Sca-1 (Sca-1$^+$). Both conclusions were actually phrased more cautiously in the reports, claiming only results consistent with their respective hypotheses.
167. Citations (Web of Science 29 May 2010): 1631 (1988), 335 (1984).
168. See Fagan (2007) for details of experimental support for the 1988 conclusion.
169. Personal communications, unless otherwise specified, refer to the qualitative interviews conducted on dates noted.
170. See Fagan (2010).
171. Today, there is no widely agreed-upon method for isolating all and only mouse HSC (Kondo 2010, 9–10, 42, Lanza et al. 2009, 153, 444–446, Wognum et al. 2003). Instead, various partly-overlapping bone marrow subsets are identified as HSC.
172. The Society for Hematology and Stem Cells (est. 1950) was incorporated in 1972 as the International Society for Experimental Hematology, with an annual meeting and monthly journal, *Experimental Hematology*. Other specialized journals soon followed: *Blood Cells* (1975–present) and *Stem Cells* (1983–present).
173. Herzenberg, pers. comm. (interview of 6 April 2007). This split paralleled a longstanding theoretical division between humoral and cellular immunology.
174. Weissman, pers. comm. (interview of 7 November 2005).
175. See Fagan (2007) for details.
176. Müller-Sieburg et al. (1986), Müller-Sieburg pers. comm. (interview of 6 April 2007), Weissman, pers. comm. (interview of 7 November 2005).
177. Ailles, pers. comm. (interview of 4 April 2007).
178. But see Chapter 9, for proposals along these lines.
179. For details, see Kitano (2002), O'Malley and Dupré (2005), and Chuang et al. (2010). More in-depth surveys include Alon (2007), Boogerd et al. (2007), Klipp et al. (2009), and Szallasi et al (2010).

180. In this chapter, I treat mathematical modeling and computer simulation together, focusing on the former's contrast with concrete experimental methods of stem cell biology. This is not to deny that mathematical modeling and computer simulation are distinct activities with interesting epistemic differences. Their contrasts, however, are not as important for understanding the relation of systems to stem cell biology, which typically uses neither.
181. For example, Kitano (2002) and Huang (2009).
182. Terms from O'Malley and Dupré (2005).
183. In this sense, systems biology includes a form of "exploratory experimentation," distinct from Steinle's sense of the term (Franklin 2005; see §7.2).
184. Accordingly, my treatment of the two fields is not symmetric, as cell development is *the* central phenomenon requiring explanation in stem cell biology.
185. Chapters 7 and 8 argue these points for stem cell biology. For systems biology, see references in note 179.
186. These include an *EMBO* conference on Stem Cells, Systems, and Synthetic Biology (2009; Cambridge, UK), Systems Biology of Stem Cells Symposium (2010; Irvine, CA), and the Harvard Stem Cell Institute's International Conference on the Systems Biology of Human Disease (2011; Cambridge, MA, USA).
187. Section 9.3 closely follows systems biologists own accounts of model-construction (e.g. Alon 2007, Klipp et al. 2009, Szallasi et al. 2010). My own proposal begins in §9.4.
188. In stochastic models, the state variable is usually molecule number, and the overall state of the system is not fully determined, but defined in terms of one or more probability distributions. For simplicity, I focus on deterministic models. The claims in this section also extend to stochastic systems models.
189. See Chapter 5.
190. Details in Alon (2007), Klipp et al. (2009), Szallasi et al. (2010), and Buchanan et al. (2010).
191. Examples of the latter include Qu and Ortoleva (2008), Enver et al. (2009), Huang et al. (2009), and Kaneko (2011).
192. In this particular model activation and inhibition of gene expression are represented as sigmoidal curves determined by Hill functions; n is the Hill coefficient and S the inflection point of the curve (details in Huang et al. 2007 and Klipp et al. 2009). This modeling framework involves a number of approximations and idealizations, notably the omission of cooperative binding.
193. The number and characteristics of steady-state solutions also depends on parameter values. Given a set of rate equations representing a molecular mechanism, however, these values are fixed. So this complication can be set aside here.
194. Some modeling frameworks in systems biology, notably GRNs, do retain Waddington's idea that genes ultimately control development (see §6.7). But this is not an assumption of systems biology in general.
195. If even one kind of adult stem cell can produce cells of many types, then there is an alternative to embryonic stem cell research that could have clinical benefits. Unsurprisingly, then, claims of adult stem cell plasticity

are continually made and considerable research effort is dedicated to testing them. See also Chapters 4 and 7.

196. Stability in models of cell development indicates whether a combination of gene expression levels can persist; it is not a thermodynamic concept.

197. For a recent survey, see Klipp et al. (2009, esp. Chapters 5 and 11).

198. Similar points hold for omics methods that use DNA–protein and protein–protein binding.

199. See Krohs and Callebaut (2007) for more on top-down and bottom-up approaches in systems biology.

200. Prominent stem cell groups that integrate omics and computational biology with traditional molecular and cellular experiments include the laboratories of Sheng Ding (Scripps), Tariq Enver (Oxford), Rudolf Jaenisch (MIT), Ihor Lemischka (Princeton), Doug Melton (Harvard), and Austin Smith (Cambridge).

201. Interestingly, because cell states are dynamic, and attainable attractors depend on initial conditions, Kauffman's derived landscape is malleable on developmental timescales.

202. See Moss (2003, 98–107) for a detailed critique.

203. For details, see Regalado (2011) and sources therein.

204. See Chapter 7 for background motivating this concern.

205. It is not so widely agreed exactly which values those are, or how they should be balanced; there is, arguably, no general answer (Kuhn 1977). For simplicity, I use "epistemic" and "cognitive" (and their contraries) interchangeably for now. The significance of the distinction (according to some) is discussed later.

206. Ethical values are treated somewhat differently (especially in the life sciences). It is widely agreed that methods of scientific inquiry should conform to ethical principles.

207. Some philosophers further distinguish between epistemic and cognitive values, the former being concerned with truth or the theory-evidence relation formally construed, and the latter with scientific aims, such as understanding and progress. On this refined classification, internal consistency and empirical accuracy are epistemic values, while external consistency, explanatory unification, simplicity, and fruitfulness are cognitive. Ethical values are classified as non-epistemic and non-cognitive. So, for simplicity, I use 'epistemic' in the wider sense.

208. This argument follows Anderson (1995).

209. In Lacey's terminology, cognitive and non-cognitive, respectively (see note above).

210. More than 30,000 HSC transplants (from bone marrow, mobilized peripheral blood, or cord blood) are performed worldwide each year to treat blood cell-related malignancies, such as leukemia. Epidermal stem cells are used to treat burns and some forms of blindness, but such treatments are not yet widely available. Other experimental stem cell treatments are still works-in-progress. See the NIH's website for a list of ongoing and planned clinical trials using stem cells in the United States.

Bibliography

Adewumi, O., Aflatoonian, B., Ahrlund-Richter, L., *et al.* (2007) 'Characterization of human embryonic stem cell lines by the International Stem Cell Initiative', *Nature Biotechnology* 25: 803–816.

Aggarwal, K. and Lee, K. (2003) 'Functional genomics and proteomics as a foundation for systems biology', *Briefings in Functional Genomics and Proteomics* 2: 175–184.

Alon, U. (2007) *An Introduction to Systems Biology.* Boca Raton: Taylor and Francis.

Anderson, E. (1995) 'Knowledge, human interests, and objectivity in feminist epistemology', *Philosophical Topics* 23: 27–58.

Andrews, P. (2002) 'From teratocarcinomas to embryonic stem cells', *Philosophical Transactions of the Royal Society of London, Series B* 357: 405–417.

Ankeny, R. (1997) 'The conqueror worm', PhD dissertation. University of Pittsburgh.

Ankeny, R. (2001) 'Model organisms as models', *Philosophy of Science* 68: S251–S261.

Bechtel, W. (2006) *Discovering Cell Mechanisms.* Cambridge: Cambridge University Press.

Bechtel, W. (2010) 'The cell: locus or object of inquiry?' *Studies in History and Philosophy of Biological and Biomedical Sciences* 41: 172–182.

Bechtel, W. and Richardson, R. (1993) *Discovering Complexity.* Princeton: Princeton University Press (2nd ed. 2010).

Bechtel, W. and Abrahamson, A. (2005) 'Explanation: a mechanist alternative', *Studies in History and Philosophy of Biological and Biomedical Sciences* 36: 421–441.

van Bekkum, D., van Noord, M. Maat, B. and Dicke, K. (1971) 'Attempts at identification of hemopoietic stem cell in mouse', *Blood* 38: 547–558.

Beurton, P., Falk, R. and Rheinberger, H.-J. (eds) (2000) *The Concept of the Gene in Development and Evolution.* Cambridge: Cambridge University Press.

Blau, H., Brazelton, T. and Weimann, J. (2001) 'The evolving concept of a stem cell', *Cell* 105: 829–841.

Bock, C., Kiskinis, E., Verstappen, G., *et al.* (2011) 'Reference maps of human ES and iPS cell variation enable high-throughput characterization of pluripotent cell lines', *Cell* 144: 1–14.

Bogen, J. and Woodward, J. (1988) 'Saving the phenomena', *Philosophical Review* 97: 303–352.

Bolker, J. (1995) 'Model organisms in developmental biology.' *BioEssays* 17: 451–455.

Boogerd, F., Bruggeman, F., Hofmeyr, J.-H. and Westerhoff, H. (eds) (2007) 'Afterthoughts as foundations for systems biology'. In: *Systems Biology: Philosophical Foundations*, pp. 321–336. Amsterdam: Elsevier.

Brandt, C. (2010) 'The metaphor of 'nuclear reprogramming'. In: Barahona, A., Suarez-Díaz, E. and Rheinberger H.-J. (eds) *The Hereditary Hourglass: Genetics and Epigenetics, 1868–2000*, pp. 85–95. Berlin: Max Planck Institute for History of Science.

Brown, N., Kraft, A. and Martin, P. (2006) 'The promissory pasts of blood stem cells', *BioSocieties* 1: 329–348.

Buchanan, M., Caldarelli, G., de los Rios, P., *et al.* (eds) (2010) *Networks in Cell Biology*. Cambridge: Cambridge University Press.

Burian, R. (2005) *The Epistemology of Development, Evolution, and Genetics*. Cambridge: Cambridge University Press.

Cai, J., Weiss, M. and Rao, M. (2004) 'In search of "stemness"', *Experimental Hematology* 32: 585–598.

Carlson, E. (1966) *The Gene – A Critical History*. New York: W.B. Saunders.

Carroll, S. (2005) *Endless Forms Most Beautiful*. New York: W.W. Norton.

Cartwright, N. (1983) *How the Laws of Physics Lie*. Oxford: Oxford University Press.

Cartwright, N. (1999) *The Dappled World*. Cambridge: Cambridge University Press.

Cartwright, N., Shomar, T. and Suárez, M. (1995) 'The tool box of science'. In: Herfel, W., Krajewski, W., Niiniluoto, I. and Wojcicki, R. (eds) *Theories and Models in Scientific Processes*, pp. 137–149. Amsterdam: Rodopi.

Cat, J. (2005) 'Modeling cracks and cracking models', *Synthese* 146: 447–487.

de Chadarevian, S. and Hopwood, N. (eds) (2004) *Models: the Third Dimension of Science*. Stanford: Stanford University Press.

Cherry, C. and Matunis, E. (2010) 'Epigenetic regulation of stem cell maintenance in the Drosophila testis via the nucleosome-remodeling factor NURF', *Cell Stem Cell* 6: 557–567.

Chuang, H.-Y., Hofree, M. and Ideker, T. (2010) 'A decade of systems biology', *Annual Review of Cell and Developmental Biology* 26: 721–744.

Clarke, A. and Fujimura, J. (eds) (1993) *The Right Tools for the Job: At Work in Twentieth-Century Life Sciences*. Princeton: Princeton University Press.

Cohen, D. and Melton, D. (2011) 'Turning straw into gold', *Nature Reviews Genetics* 12: 243–252.

Conrad, E. and Tyson, J. (2010) 'Modeling molecular interaction networks with nonlinear ordinary differential equations'. In: Szallasi, Z., Stelling, J. and Periwal, V. (eds) *Systems Modeling in Cell Biology*, pp. 97–123. Cambridge: MIT Press.

Craver, C. (2007) *Explaining the Brain*. Oxford: Oxford University Press.

Craver, C. and Bechtel, W. (2007) 'Top-down causation without top-down causes', *Biology and Philosophy* 22: 547–563.

Creager, A. (2002) *The Life of a Virus*. Chicago: University of Chicago Press.

Creager, A., Lunbeck, E. and Wise, M.N. (eds) (2007) *Science Without Laws*. Durham: Duke University Press.

Darden, L. (2006) *Reasoning in Biological Discoveries*. Cambridge: Cambridge University Press.

Davidson, E. (2006) *The Regulatory Genome*. Academic Press, Elsevier.

Davidson, E. and Levine, M. (2008) 'Properties of developmental gene regulatory networks', *Proceedings of the National Academy of Sciences USA* 105: 20063–20066.

Douglas, H. (2009) *Science, Policy, and the Value-Free Ideal*. Pittsburgh: University of Pittsburgh Press.

Doulatov, S., Notta, F., Laurenti, E. and Dick, J. (2012) 'Hematopoiesis: a human perspective', *Cell Stem Cell* 10: 120–136.

Ema, H., Kobayashi, T. and Nakauchi, H. (2010) 'Principles of hematopoietic stem cell biology'. In: Kondo, M. (ed.) *Hematopoietic Stem Cell Biology*, pp. 1–36. New York: Humana Press.

Enver, T., Pera, M. Peterson, C. and Andrews, P. (2009) 'Stem cell states, fates, and the rules of attraction', *Cell Stem Cell* 4: 387–397.

European Stem Cell Network (2008) 'Stem cell glossary', available at http://www.eurostemcell.org/stem-cell-glossary#letters (accessed February 2011).

Evans, M. and Kaufman, M. (1981) 'Establishment in culture of pluripotential cells from mouse embryos', *Nature* 292: 154–156.

Fagan, M.B. (2007) 'The search for the hematopoietic stem cell', *Studies in History and Philosophy of Biological and Biomedical Sciences* 38: 217–237.

Fagan, M.B. (2010) 'Stems and standards', *Journal of the History of Biology* 43: 67–109.

Fagan, M.B. (2011) 'Social experiments in stem cell biology', *Perspectives on Science* 19: 235–262.

Fagan, M.B. (2012) 'Waddington redux', *Biology and Philosophy* 27: 179–213.

Fehr, C. and Plaisance, K. (2010) 'Making philosophy of science more socially relevant', *Synthese* 177: 301–492.

Fortunel, N., Otu, H.H., Ng, H.H., *et al.* (2003) 'Comment on "'stemness': transcriptional profiling of embryonic and adult stem cells" and a "stem cell molecular signature"', *Science* 302: 393.

Fox, C. (2006) *Cell of Cells*. New York: W.W. Norton.

Frank, S. (1995) 'Price's contributions to evolutionary genetics', *Journal of Theoretical Biology* 175: 373–388.

Franklin, A. (1986) *The Neglect of Experiment*. Cambridge: Cambridge University Press.

Franklin, L. (2005) 'Exploratory experiments', *Philosophy of Science* 72: 888–899.

van Fraassen, B. (1989) *Laws and Symmetry*. Oxford: Clarendon.

French, S. and Ladyman, J. (1999) 'Reinflating the semantic approach', *International Studies in Philosophy of Science* 13: 103–121.

Galison, P. (1987) *How Experiments End*. Oxford: Oxford University Press.

Gibson, D., Glass, J.I., Lartigue, C., *et al.* (2010) 'Creation of a bacterial cell controlled by a chemically-synthesized genome', *Science* 329: 52–56.

Giere, R. (1988) *Explaining Science*. Chicago: Chicago University Press.

Giere, R. (2004) 'How models are used to represent reality', *Philosophy of Science* 71: 742–752.

Gilbert, S. (1991) 'Epigenetic landscaping: Waddington's use of cell fate bifurcation diagrams', *Biology and Philosophy* 6: 135–154.

Gilbert, S. (2001) 'Ecological developmental biology', *Developmental Biology* 233: 1–12.

Glennan, S. (1996) 'Mechanisms and the nature of causation', *Erkenntnis* 44: 49–71.

Glennan, S. (2002) 'Rethinking mechanistic explanation', *Philosophy of Science* 69: S342–S353.

Gottweis, H., Salter, B. and Waldby, C. (2009) *The Global Politics of Human Embryonic Stem Cell Science*. Basingstoke: Palgrave MacMillan.

Griffiths, P. and Stotz, K. (2007) 'Gene'. In: Hull, D. and Ruse, M. (eds) *The Cambridge Companion to the Philosophy of Biology*, pp. 85–102. Cambridge: Cambridge University Press.

Hacking, I. (1983) *Representing and intervening*. Cambridge: Cambridge University Press.

Hacking, I. (1992) 'The self-vindication of the laboratory sciences'. In: Pickering, A. (ed.) *Science as Practice and Culture*, pp. 29–64. Chicago: University of Chicago Press.

Haraway, D. (1976) *Crystals, Fabrics, and Fields*. New Haven: Yale University Press.

Hitchcock, C. and Woodward, J. (2003) 'Explanatory Generalizations, Part II.' *Noûs* 37: 181–199.

Hochedlinger, K. and Plath, K. (2009) 'Epigenetic reprogramming and induced pluripotency', *Development* 136: 509–523.

Huang, S. (2009) 'Reprogramming cell fates.' *BioEssays* 31: 546–560.

Huang, S. (2010) Conference presentation on 'Systems Biology of Stem Cells.' University of California, Irvine.

Huang, S., Guo Y.-P., May, G. and Enver, T. (2007) 'Bifurcation dynamics in lineage-commitment in biopotent progenitor cells', *Developmental Biology* 305: 695–713.

Huang, S., Ernberg, I. and Kauffman, S. (2009) 'Cancer attractors', *Seminars in Cell and Developmental Biology* 20: 869–876.

Hughes, R.I.G. (1997) 'Models and representation', *Philosophy of Science* 64: S325–S336.

Ichida, J., Kiskinis, E. and Eggan, K. (2010) 'Shushing down the epigenetic landscape towards stem cell differentiation.' *Development* 137: 2455–2460.

Illari, P.M.K. and Williamson, J. (2010) 'Function and organization', *Studies in History and Philosophy of Biological and Biomedical Sciences* 41: 279–291.

Jablonka, E. and Lamb, M. (2005) *Evolution in Four Dimensions*. Cambridge: MIT Press.

Jacob, F. (1978) 'Mouse teratocarcinoma and mouse embryo', *Proceedings of the Royal Society of London, Series B* 201: 249–270.

Kaneko, K. (2011) 'Characterization of stem cells and cancer cells on the basis of gene expression profile stability, plasiticity, and robustness', *BioEssays* 33: 403–413.

Kauffman, S. (1969) 'Metabolic stability and epigenesis in randomly constructed genetic nets', *Theoretical Biology* 22: 437–467.

Kauffman, S. (1993) *The Origins of Order*. Oxford: Oxford University Press.

Keating, P. and Cambrosio, A. (2003) *Biomedical Platforms*. Cambridge: MIT Press.

Keller, E.F. (2002) *Making Sense of Life*. Cambridge: Harvard University Press.

Kimmelman, B. (1993) 'Organisms and interests in scientific research'. In: Clarke, A. and Fujimura, J. (eds) *The Right Tools for the Job*, pp. 198–232. Princeton: Princeton University Press.

Kincaid, H., Dupré, J. and Wylie, A. (eds) (2007) *Value-free Science?* Oxford: Oxford University Press.

Kitano, H. (2002) 'Systems biology: a brief overview', *Science* 295: 1662–1664.

Kitcher, P. (1981) Explanatory unification. *Philosophy of Science* 48: 507–531.

Kitcher, P. (2010) *Science in a Democratic Society*. New York: Prometheus Books.

Kleinsmith, L. and Pierce, G.B. (1964) 'Multipotentiality of single embryonal carcinoma cells', *Cancer Research* 24: 1544–1551.

Klipp, E., Liebermeister, W., Wierling, C., *et al.* (2009) *Systems Biology: A Textbook.* Weinheim: Wiley-VCH.

Kohler, R. (1994) *The Lords of the Fly.* Chicago: University of Chicago Press.

Kondo, M. (ed.) (2010) *Hematopoietic Stem Cell Biology.* New York: Humana Press.

Koretzky, G. and Monroe, J. (2002) 'Introduction'. *Immunological Reviews* 185: 5–6.

Kourany, J. (2010) *Philosophy of Science After Feminism.* Oxford: Oxford University Press.

Kraft, A. (2009) 'Manhattan transfer: Lethal radiation, bone marrow transplantation, and the birth of stem cell biology, 1942–1961', *Historical Studies in the Natural Sciences* 39: 171–218.

Kraft, D. and Weissman, I.L. (2005) 'Hematopoietic stem cells: basic science to clinical applications', In: Bongso, A. and Lee, E.H. (eds) *Stem Cells: From Benchtop to Bedside*, pp. 253–292. Singapore: World Scientific Publishing Company.

Krohs, U. and Callebaut, W. (2007) 'Data without models merging with models without data'. In: Boogerd, F., Bruggeman, F.J., Hofmeyr, J.-H.S. and Westerhoff, H.V. (eds.) *Systems Biology: Philosophical Foundations*, pp. 181–213. Amsterdam: Elsevier.

Kuhn, T. (1977) 'Objectivity, values and theory choice'. In: *The Essential Tension*, pp. 320–339. Chicago: University of Chicago Press.

Lacey, H. (2005) 'On the interplay of the cognitive and the social in scientific practices', *Philosophy of Science* 72: 977–988.

Landecker, H. (2007) *Culturing Life.* Cambridge: Harvard University Press.

Lander, A. (2009) 'The 'stem cell' concept: is it holding us back?' *Journal of Biology* 8: 70.1–70.6.

Lanza, R., Gearhart, J., Hogan, B., *et al.* (2009) (eds) *Essentials of Stem Cell Biology*, 2nd edn. San Diego: Academic Press.

Leychkis, Y., Munzer, S. and Richardson, J. (2009) 'What is stemness?' *Studies in History and Philosophy of Biological and Biomedical Sciences* 40: 312–320.

Lillie, F. (1927) 'The gene and the ontogenetic process', *Science* 66: 361–368.

Loeffler, M. and Potten, C. (1997), 'Stem cells and cellular pedigrees: a conceptual introduction'. In: Potten, C. (ed.) *Stem Cells*, pp. 1–27. London: Academic Press.

Longino, H. (1995) 'Gender, politics, and the theoretical virtues', *Synthese* 104: 383–397.

Longino, H. (2002) *The Fate of Knowledge.* Princeton: Princeton University Press.

Lord, B. and Dexter, T.M. (1988) 'Purification of haemopoietic stem cells – the end of the road?' *Immunology Today* 9: 376–377.

Loring, J. and Rao, M. (2006) 'Establishing standards for the characterization of human embryonic stem cell lines', *Stem Cells* 24: 145–150.

Löwy, I. and Gaudillére, C. (1998) 'Disciplining cancer: mice and the practice of genetic purity'. In: Gaudillére, C. and Löwy, I. (eds) *The Invisible Industrialist* pp. 209–249. London: MacMillan.

Machamer, P., Darden, L. and Craver, C. (2000) 'Thinking about mechanisms', *Philosophy of Science* 67: 1–25.

Maehle, A-H. (2011) 'Ambiguous cells', *Notes and Records of the Royal Society* 65: 359–378.

Maherali, N. and Hochedlinger, K. (2008) 'Guidelines and techniques for the generation of induced pluripotent stem cells', *Cell Stem Cell* 3: 595–605.

Maienschein, J. (2003) *Whose View of Life?* Cambridge: Harvard University Press.

Martin, G. (1981) 'Isolation of a pluripotent cell line from early mouse embryos cultured in a medium conditioned by teratocarcinoma stem cells', *Proceedings of the National Academy of Sciences, USA* 78: 7634–7638.

Martin, P., Brown, N. and Kraft, A. (2008), 'From bedside to bench?' *Science as Culture* 17: 29–41.

Mayo, D. (1996) *Error and the growth of experimental knowledge.* Chicago: University of Chicago Press.

Melton, D. and Cowan, C. (2009) 'Stemness: definitions, criteria, and standards'. In: Lanza, R., Gearhart, J., Hogan, B. and Melton, D. (eds) *Essentials of Stem Biology*, 2nd edn, pp. xxii–xxix. San Diego: Academic Press.

Mikkers, H. and Frisén, J. (2005) 'Deconstructing stemness', *EMBO Journal* 24: 2715–2719.

Mintz, B. and Illmensee, K. (1975) 'Normal genetically mosaic mice produced from malignant teratocarcinoma cells', *Proceedings of the National Academy of Sciences, USA* 72: 3585–3589.

Mitman, G. and Fausto-Sterling, A. (1993) 'Whatever happened to *Planaria*?'. In: Clarke and Fujimura (eds) *The Right Tools for the Job*, pp. 172–197. Princeton: Princeton University Press.

Monroe, K., Miller, R. and Tobis, J. (2008) (eds) *Fundamentals of the Stem Cell Debate*. Berkeley: University of California Press.

Moreira, T. and Palladino, P. (2005) 'Between truth and hope', *History of the Human Sciences* 18: 55–82.

Morgan, M. and Morrison, M. (eds) (1999), *Models as Mediators*. Cambridge: Cambridge University Press.

Morgan, M. (2007) 'Afterword'. In: Creager, A., Lunbeck, E. and Wise, M.N. (eds) *Science Without Laws*, pp. 264–274. Durham: Duke University Press.

Morgan, T.H. (1926) *The Theory of the Gene*. New Haven: Yale University Press.

Morselt, A., Leene, W. and Visser, J. (1979) 'A cytophotometric and flow cytofluorometric approach to the differentiation of T lymphocytes in the thymus', *Histochemistry and Cell Biology* 62: 65–76.

Moss, L. (2003) *What Genes Can't Do*. Cambridge: MIT Press.

Müller-Sieburg, C., Whitlock, C. and Weissman, I.L. (1986) 'Isolation of two early B lymphocyte progenitors from mouse marrow', *Cell* 44: 653–662.

Nadir, A. (2006), 'From the atom to the cell', *Stem Cells and Development* 15: 488–491.

NIH (National Institutes of Health) (2008) 'Translational research', available at http://commonfund.nih.gov/clinicalresearch/overview-translational.aspx (accessed 29 February 2012).

NIH (National Institutes of Health) (2009) 'Stem cell basics'. In: *Stem Cell Information*, available at http://stemcells.nih.gov/info/basics/defaultpage (accessed February 2011).

Noble, D. (2010) 'Biophysics and systems biology', *Philosophical Transactions of the Royal Society A*. 368: 1125–1139.

Okasha, S. (2006) *Evolution and the Levels of Selection*. Oxford: Oxford University Press.

O'Malley, M. and Dupré, J. (2005) 'Fundamental issues in systems biology', *BioEssays* 27: 1270–1276.

Plath, K. and Lowry, W. (2011) 'Progress in understanding reprogramming to the induced pluripotent state', *Nature* 12: 253–262.

Potten, C. (ed.) (1983) 'Epithelial proliferative subpopulations'. In: *Stem Cells*, pp. 317–334. Edinburgh: Churchill-Livingstone [2nd edn (1997) London: Academic Press].

Potten, C. and Loeffler, M. (1990) 'Stem cells: attributes, cycles, spirals, pitfalls and uncertainties', *Development* 110: 1001–1020.

Price, G. (1970) 'Selection and covariance', *Nature* 227: 520–521.

Price, G. (1972) 'Extension of selection and covariance mathematics', *Annals of Human Genetics* 35: 485–490.

Price, G. (1995) 'The nature of selection', *Journal of Theoretical Biology* 175: 389–396.

Quesenberry, P., Colvin, G. and Lambert, J-F. (2002) 'The chiaroscuro stem cell', *Blood* 100: 4266–4271.

Qu, K. and Ortoleva, P. (2008) Understanding stem cell differentiation through self-organization theory. *Journal of Theoretical Biology* 250: 606–620.

Rader, K. (2004) *Making Mice*. Princeton: Princeton University Press.

Radetsky, P. (1995) 'The mother of all blood cells', *Discover* 16: 86–93.

Ramalho-Santos, M. and Willenbring, H. (2007) 'On the origin of the term "stem cell"', *Cell Stem Cell* 1: 35–38.

Regalado, A. (2011) 'Stem-cell gamble', *Technology Review* 114: 52–59.

Rheinberger, H.J. (1997) *Toward a history of epistemic things*. Stanford: Stanford University Press.

de los Rios, P. and Vendruscolo, M. (2010) 'Network views of the cell'. In: Buchanan, M., Caldarelli, G., de los Rios, P., *et al.* (eds) *Networks in Cell Biology*, pp. 4–13. Cambridge: Cambridge University Press.

Robert, J. (2005) 'Model systems in stem cell biology', *BioEssays* 26: 1005–1012.

Rose, N. (2007) *The Politics of Life Itself*. Princeton: Princeton University Press.

Salmon, W. (1989) *Four Decades of Scientific Explanation*. Minneapolis: University of Minnesota Press.

Scadden, D. and Srivastavra, A. (2012) 'Advancing stem cell biology toward stem cell therapeutics', *Cell Stem Cell* 10: 149–150.

Schaffner, K. (1993) *Discovery and Explanation in Biology and Medicine*. Chicago: University of Chicago Press.

Schofield, R. (1978) 'The relationship between the spleen colony-forming cell and the haemopoietic stem cell', *Blood Cells* 4: 7–25.

Shamblott, M., Axelman, J., Wang, S., *et al.* (1998) 'Derivation of pluripotent stem cells from cultured human primordial germ cells', *Proceedings of the National Academy of Sciences USA* 95: 13726–13731.

Sheng, X., Posenau, T., Gumulak-Smith, J., *et al.* (2009) 'Jak-STAT regulation of male germline stem cell establishment during *Drosophila* embryogenesis', *Developmental Biology* 334: 335–344.

Shostak, S. (2006) '(Re)defining stem cells', *BioEssays* 28: 301–308.

Simmons, P. (2007) 'The ISSCR: moving the field forward together', *Cell Stem Cell* 1: 51–52.

Sober, E. (2008) *Evidence and Evolution*. Cambridge: Cambridge University Press.

Spangrude, G. (1989) 'Enrichment of murine hematopoietic stem-cells: diverging roads', *Immunology Today* 10: 344–350.

Spangrude, G. and Slayton, W. (2009) 'Isolation and characterization of hematopoietic stem cells'. In: Lanza, R., Gearhart, J., Hogan, B. and Melton, D. (eds) *Essentials of Stem Cell Biology*, 2nd edn, pp. 438–443. San Diego: Academic Press.

Spangrude, G., Heimfeld, S. and Weissman, I.L. (1988) 'Purification and characterization of mouse hematopoietic stem cells', *Science* 241: 58–62.

Stadtfeld, M. and Hochedlinger, K. (2010) 'Induced pluripotency: history, mechanisms, and applications', *Genes and Development* 24: 2239–2263.

Steinle, F. (2002) 'Experiments in history and philosophy of science', *Perspectives on Science* 10: 408–432.

Stevens, L.C. and Little, C.C. (1954) 'Spontaneous testicular teratomas in an inbred strain of mice.' *Proceedings of the National Academy of Sciences USA* 40: 1080–1087.

Suárez, M. (2004) 'An inferentialist conception of scientific representation', *Philosophy of Science* 71: 767–779.

Suárez, M. (ed.) (2008) *Fictions in Science*. New York: Routledge.

Suárez, M. and Cartwright, N. (2008), 'Theories: tools versus models', *Studies in History and Philosophy of Modern Physics* 39: 62–81.

Szallasi, Z., Stelling, J. and Periwal, V. (eds) (2010) *Systems Modeling in Cell Biology*. Cambridge: MIT Press.

Takahashi, K. and Yamanaka, S. (2006) 'Induction of pluripotent stem cells from mouse embryonic and adult fibroblast cultures by defined factors', *Cell* 126: 663–676.

Takahashi, K., Tanabe, K., Ohnuki, M., *et al.* (2007) 'Induction of pluripotent stem cells from adult human fibroblasts by defined factors', *Cell* 131: 861–872.

Thomas, K. and Capecchi, M. (1987) 'Site-directed mutagenesis by gene targeting in mouse embryo-derived stem cells', *Cell* 51: 503–512.

Thomson, J., Itskovitz-Eldor, J., Shapiro, S., *et al.* (1998) 'Embryonic stem cell lines derived from human blastocysts', *Science* 282: 1145–1147.

Till, J. and McCulloch, E. (1961) 'A direct measurement of the radiation sensitivity of normal mouse bone marrow cells', *Radiation Research* 14: 213–222.

Till, J.E., McCullough, E.A. and Siminovitch, L. (1964), 'A stochastic model of stem cell proliferation, based on the growth of spleen colony-forming cells', *Proceedings of the National Academy of Sciences USA* 51: 29–36.

Trounson, A. (2009) 'Why stem cell research'. In: Lanza, R., Gearhart, J., Hogan, B. and Melton, D. (eds) *Essentials of Stem Cell Biology*, 2nd edn, p. xix. San Diego: Academic Press.

Van Speybrock, L., Van de Vuver, G. and De Waele, D. (eds) (2002) *From Epigenesis to Epigenetics*. Annals of the New York Academy of Sciences, Volume 981.

Visser, J. and de Vries, P. (1988) 'Isolation of spleen-colony forming cells (CFU-s) using wheat germ agglutinin and rhodamine 123 labeling', *Blood Cells* 14: 369–384.

Visser, J. and de Vries, P. (1994) 'Analysis and sorting of hematopoietic stem cells from mouse bone marrow', *Methods in Cell Biology* 42: 243–261.

Visser, J., van den Engh, G. and van Bekkum, D. (1980) 'Light-scattering properties of murine hemopoietic cells', *Blood Cells* 6: 391–407.

Visser, J., Bauman, J., Mulder, A., *et al.* (1984) 'Isolation of murine pluripotent hemopoietic stem cells', *Journal of Experimental Medicine* 59: 1576–1590.

Vogel, H., Niewisch, H. and Matioli, G. (1969) 'Stochastic development of stem cells', *Journal of Theoretical Biology* 22: 249–270.

Waddington, C.H. (1939) *An Introduction to Modern Genetics.* New York: MacMillan.

Waddington, C.H. (1940) *Organisers and Genes.* Cambridge: Cambridge University Press.

Waddington, C.H. (1956) *Principles of Embryology.* New York: MacMillan.

Waddington, C.H. (1957) *The Strategy of the Genes.* London: Taylor & Francis.

Wang, W. and Audet, J. (2009) 'Single-cell approaches to dissect cellular signaling networks'. In: Rajasekhar, V. and Vemuri, M. (eds) *Regulatory Networks in Stem Cells,* pp. 337–345. New York: Humana Press.

Waters, C.K. (2007) 'Causes that make a difference', *Journal of Philosophy* 104: 551–579.

Watt, F. and Driskell, R. (2010) 'The therapeutic potential of stem cells', *Philosophical Transactions of the Royal Society, Series B* 365: 155–163.

Weber, M. (2005) *The Philosophy of Experimental Biology.* Cambridge: Cambridge University Press.

Weber, M. (2007) 'Redesigning the fruit fly'. In: Creager, A., Lunbeck, E. and Wise, M. N. (eds) (2007) *Science Without Laws* pp. 23–45. Durham: Duke University Press.

Williams, C., Wainwright, S. P., Ehrich, K. and Michael, M. (2008) 'Human embryos as boundary objects?' *New Genetics and Society* 27: 7–18.

Wimsatt, W. (2007) *Re-engineering Philosophy for Limited Beings.* Cambridge: Harvard University Press.

Wognum, A., Eaves, A.C. and Thomas, T. (2003) 'Identification and isolation of hematopoietic stem cells', *Archives of Medical Research* 34: 461–475.

Woodward, J. (2002) 'What is a mechanism? A counterfactual account', *Philosophy of Science,* 69: S366–S377.

Woodward, J. (2003) *Making Things Happen.* Oxford: Oxford University Press.

Woodward, J. (2010) 'Causation in biology', *Biology and Philosophy* 25: 287–318.

Woodward, J., and Hitchcock, C. (2003) Explanatory generalizations, Part I. *Noûs* 37: 1–24.

Yamanaka, S. (2007) 'Strategies and new developments in the generation of patient-specific pluripotent stem cells', *Cell Stem Cell* 1: 39–49.

Yamanaka, S. (2009) 'Elite and stochastic models for induced pluripotent stem cell generation', *Nature* 460: 49–52.

Yamashita, Y., Jones, D. and Fuller, M. (2003) 'Orientation of asymmetric stem cell division by the APC tumor suppressor and centrosome', *Science* 301: 1547–1550.

Zhou, Q. and Melton, D. (2008) 'Extreme makeover', *Cell Stem Cell* 3: 382–388.

Zipori, D. (2004) 'The nature of stem cells', *Nature Reviews Genetics* 5: 873–878.

Zipori, D. (2005) 'The stem state', *Stem Cells* 23: 719–726.

Zipori, D. (2009) *Biology of Stem Cells and the Molecular Basis of the Stem State.* New York: Humana Press.

Index

Printed in the United States
by Baker & Taylor Publisher Services